Cultivated Building Materials

Layout, cover design and typesetting
Miriam Bussmann

Translation from German into English
Margot Stringer (Contributions by Läpple and Moravánszky)

Production
Katja Jaeger

Paper
135g Hello Fat matt 1.1

Printing
Grafisches Centrum Cuno GmbH & Co. KG, Calbe

Library of Congress Cataloging-in-Publication data
A CIP catalog record for this book has been applied for at the Library of Congress.

Bibliographic information published by the German National Library
The German National Library lists this publication in the Deutsche Nationalbibliografie;
detailed bibliographic data are available on the Internet at http://dnb.dnb.de.

This publication is also available as an e-book (ISBN PDF 978-3-0356-0892-2)

© 2017 Birkhäuser Verlag GmbH, Basel
P.O. Box 44, 4009 Basel, Switzerland
Part of Walter de Gruyter GmbH, Berlin/Boston

Printed on acid-free paper produced from chlorine-free pulp. TCF ∞
Printed in Germany

ISBN 978-3-0356-1106-9

9 8 7 6 5 4 3 2 1

www.birkhauser.com

CONTENTS

Cultivated Building Materials

CHALLENGES, STRATEGIES, AND GOALS

INTRODUCTION: CULTIVATED BUILDING MATERIALS

Dirk E. Hebel and Felix Heisel

The 21st century will face a radical paradigm shift in how we produce materials for the construction of our habitat. While the period of the first industrial revolution, in the 18th and 19th century has resulted in a conversion from regenerative (agrarian) to non-regenerative material sources (mines), our time might experience the reverse: a shift towards cultivating, breeding, raising, farming, or growing future resources going hand in hand with a reorientation of biological production methods and goals.

While agricultural products so far are mostly associated with the food industry, this book wants to explore the option of biological products as a major resource for the building industry, in particular within the processes of architectural design and production. Such nurtured construction resources can be cultivated either within the conventional soil-based agricultural framework or in purpose-built breeding farms, using micro-organisms that until very recently were not even considered useful for the building industry. A third option is the adaptation of biological growing patterns and procedures, multiplying natural phenomena within a controlled environment determined by design parameters. In our view, while all three scenarios are important, the latter two also respond elegantly to the current trend towards small-scale industrial and agricultural production units within urban environments, as Dieter Läpple points out convincingly and encouragingly in his contribution (see p. 20).

Addressing the industrialization of building materials, we are aware of the fact that the term *industrialization* describes different periods in history. The first industrial revolution, starting in the 18th century and reaching its climax in the 19th century, was based on the invention of the steam engine to mechanize production. The second industrial revolution is described by the use of electrical power engines to initiate mass production, starting at the end of the 19th century. The third industrial revolution, commencing in the middle of the 20th century, has used digital information technology to automate production. A fourth industrial revolution could lie just ahead of us, reintroducing biological aspects into an otherwise often mechanized industrialized world – "characterized by a fusion of technologies that is blurring the lines between the physical, digital, and biological spheres", in the words of Klaus Schwab, Founder and Executive Chairman of the World Economic Forum.[1]

Status quo: standardization

When considering the use of natural organic raw materials for the building industry, biological imperfections are going against our current understanding of the internationally accepted norms of standardized mass production. Therefore, next to a continuous interest in advancing material properties, efforts of mass production of biological materials face the need to normalize nature in order to guarantee predictable and controllable properties. This need for globally accepted norms and standards emerged with industrialization. Peter Berz, in his publication *08/15 – a standard of the 20th century,*[2] provides a detailed and elucidating view of the historical development of normative systems. The concept of standardization originates in the fact that any machine is composed of a multitude of parts, which need maintenance and sometimes replacement. In early machine-like apparatuses, such as the printing press invented in Germany by Gutenberg in the 15th century, pieces would be replaced without referring to an absolute measurement system, instead following a dialectical system where a piece was remodelled in response to an existing component until both elements fit each other. In this kind of self-referencing system, parts could only be produced one after the other, referencing their neighbouring element.

Modularity, by contrast, as a technical principle uses an absolute reference system that is not physically present in the machine itself. This system is normative: elements are modular because they are identical with a normative measurement and referenced outside their physical existence and context. This production principle, which can be described as mensuration, allows the exchange of components based on precise dimensions defined in technical drawings.

Scales, rulers, and guides entered the world of mechanical production at the beginning of the 19th century; but along with them went individual errors in taking and transferring measurements by hand that quickly revealed the limits of this system. In the context of mass weapon production, where innovations were most urgently enforced by governments worldwide, the American Simeon North developed a completely new mechanical process of scale-reference. He introduced the idea to produce for each machine part its own checking device: a solid measure or gauge. This principle of a mirrored physiognomy or mirrored reference was adapted in 1817 by all state weapon manufacturers in the USA. Each component was reworked and shaped until it

fit into the gauge. Producers were given a set of gauges in order to guarantee identical products. Although the copies of the original gauge underwent regular checking procedures in the manufactories, problems occurred through abrasion and imperfection in the production processes. The gauge system was still a hybrid between a self-referencing system and an absolute reference system.[3]

Only the introduction of machines producing normative components for normative machines was able to overcome this problem. In an additional step, the introduction of a technical measuring process based on limited tolerances and the so-called go/no go gauge further advanced this methodology. Inventor of this new understanding of standardization was Georg Schlesinger, who, as an employee of the Loewe Factory in Berlin, Germany, actually created a new industrial field: machine building. In 1904, he became the first chairholder of *Tool-making Machines and Science of Management* at the Technical University of Charlottenburg (Berlin), the first professorship of its kind in Europe.[4] Another protagonist of an increasing standardization process in Germany, Werner von Siemens, co-founded the first *Electrotechnical Society* (Elektrotechnischer Verein) in 1879. It was the first association worldwide aiming for a standardized development of technical applications of electricity. Uniting several other associations formed in the following years, the *Association of German Electrical Engineers* (Verband Deutscher Elektrotechniker) was founded in 1893. This union established the first norm, by the title of *Safety Regulations for Electrical Heavy Current Installations* (Sicherheitsvorschriften für elektrische Stark-

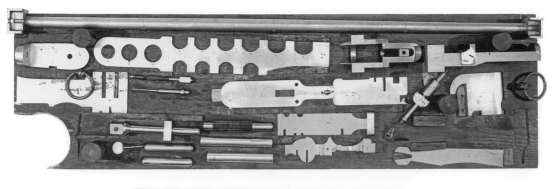

stromanlagen), in 1895. Next a national association addressing all issues of the industrial process was formed in 1917, when in the midst of the technologically dominated First World War the German Reich wanted to rationalize its arms industry by forming a coherent production process to provide the army with standardized weapons. To this purpose the *Royal Fabrication Bureau for Artillery* (Königliches Fabrikationsbüro für Artillerie) in Berlin aimed for a national norm especially addressing the young industry of machine building, so that all manufacturers would serve only one client: the German army. The *Standards Committee for Mechanical Engineering* (Normalienausschuss für den Maschinenbau) was founded in May 1917 to stan-

dardize the most important machine components within the German Reich. It was renamed after few months to *Standardization Committee of German Industry* (Normenausschuß der deutschen Industrie) so as to address an even wider field of industrial production. The results of this regulative process were published under the name of *German Industry Standard* (Deutsche Industrienorm) or the abbreviation DI-Norm, later to become DIN. The presumably best-known DIN, addressing formats of paper, like DIN A4, was established in 1922.

The norms defined by the numerous national committees and associations founded in the first half of the 20th century were incompatible to each

▲ **The Springfield Standard represents a set of 11 gauges that was sent to all factories of the United States standard rifle in 1923, in order to check elements during production.**

▾ **A car of the era of the third industrial revolution consists of several thousand single, exchangeable pieces. Student exercise at the Assistant Professorship Dirk E. Hebel at the ETH Zürich in 2016, aimed at disassembling a Golf 3 station wagon.**

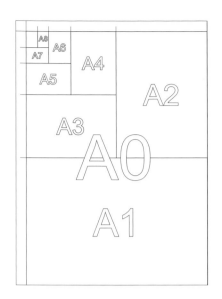

◄ **The DIN paper formats were introduced in 1922.**

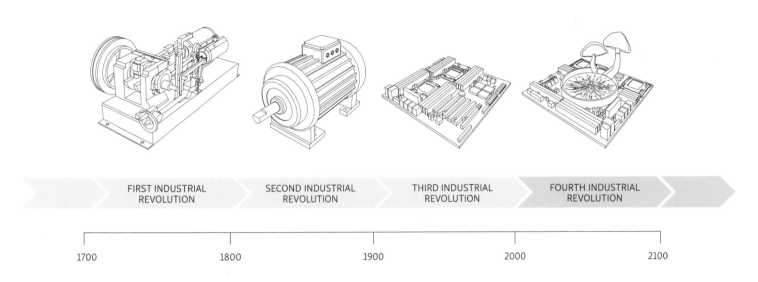

| FIRST INDUSTRIAL REVOLUTION | SECOND INDUSTRIAL REVOLUTION | THIRD INDUSTRIAL REVOLUTION | FOURTH INDUSTRIAL REVOLUTION |

1700   1800   1900   2000   2100

other and created increasing problems within a growing international industrial market. Therefore, in 1946, an international conference of 25 national institutes agreed to establish the International Organization for Standardization (ISO), with the headquarters to be based in Geneva, Switzerland. Today (as of August 2016), 163 nations are members of ISO.

The fourth industrial revolution – a new order

While the mechanized world that has emerged since the first industrial revolution aims to establish an endlessly reproducible state of "exactly the same", any biologically-driven process resembles a unique process in itself, allowing an endless variety of "almost the same". The digital may be seen as an in-between state. Setting – at least for a while – the physical dimensions of a possible product aside and, for example, calculating or mimicking a growth process with its expected variations in order to find the best circumstances, can help generate and cultivate particular adaptive qualities. In this way, the digital realm can be used to plan the sourcing of the best growing conditions from available data pools without undergoing a time-consuming trial-and-error process. Here, a new understanding of a bio-mechanized or bio-industrial production is required, which allows for a higher degree of deviations and abnormalities. This mentality would correspond with the theories of Georges Canguilhem, who introduced the idea of "another" normality, or the state of an accepted pathology, in his publication *The Normal and the Pathological*.[5] While an organism can adapt to a crisis by accepting a new state as another kind of

▲ **An overview of the four industrial revolutions from the 18th century until today.**

normality (Canguilhem uses among others the example of amputation), a machine is unable to compensate for the smallest deviation from the normal state for which it was designed, threatening dysfunction, inefficiency or damage. Therefore, Canguilhem defines the machine as such by its capacity to be dissembled into or assembled with standardized or normed elements which, through their physical form, interact with each other to reach a state of a single normality.

At the other extreme, a biological organism features a multiplicity of normalities and is able to switch between them. Nature does not strive for a constant norm, but instead establishes a constant normative process, where varieties of normal states can exist next to each other. Canguilhem describes this phenomenon of different and simultaneous norms by comparing it to the relationship between health (usually understood as a norm) and disease (usually understood as an abnormality). The normal (body), or to speak in this analogy: the healthy (body) has the capacity to adapt constantly to new situations and conditions. When it falls sick (becomes abnormal) it is losing this ability. "The sick living being is normalized in well-defined conditions of existence and has lost his normative capacity, the capacity to establish other norms in other conditions."[6] "The healthy organism tries less to maintain itself in its present state and environment than to realize its nature. This requires that the organism, in facing risks, accepts the eventuality of catastrophic reactions. The healthy man (...) measures his health in terms of his capacity to overcome organic crises in order to establish a new order."[7] "What characterizes health is the possi-

bility of transcending the norm, which defines the momentary normal, the possibility of tolerating infractions of the habitual norm and instituting new norms in new situations."[8]

Seen in this light, the idea of using biologically cultivated materials for an industrial production process, controlled by a system of norms as it is established in our society, encounters its greatest challenge. In the following chapters of this publication, Christoph Ziegert, Jasmine Alia Blaschek, Pablo van der Lugt, Nikita Aigner, and Philipp Müller describe the discussions and endeavours for a number of materials in the sometimes impossible quest for the single appropriate norm which our laws and regulations tend to require.

Any such shift towards a new order in the industrialization of building materials needs to respect, adapt, and further develop the established frameworks of norms, regulations, and standards concerning safety, health, and durability in order to be operational. The concept of industrialized building with nature does not promote a move backwards to the pre-industrial age. The present publication seeks to describe ways to progress within the given industrial setting in order to modify and reinvent it. This also applies to the social conditions of production and the cultural acceptance of new materials and construction methods. The introduction of biologically driven and digitally supported production processes could allow so-called late-comer economies in less industrialized areas of our planet to participate in this revolution. A similar thought was previously formulated by Thorstein Veblen in his thesis of 1919,[9] addressing Germany and Japan as latecomers to global industrialism. From this perspective, latecomers have the advantage of being able to use the knowledge and advancements of further developed nations without having to go through the same (painful) period of their elaboration. "This privilege encompasses all technological and organizational innovations. The challenge, obviously, is to identify the good in both the new and the existing and blend them with a minimum negative consequence."[10]

Cultivated building materials, considered in the context of a possible fourth industrial revolution, could constitute a very promising element of a new economic, ecological, social, and cultural order – which, this time, might emerge from regions of the Global South and add a new chapter to the history of industrial production, with protagonists as ingenious and far-reaching as those of the 19th and 20th century.

The mining mentality

The process of industrialization saw the shift from cultivating to mining in the 18th century. As of today, steel-reinforced concrete is the most used construction material produced on an industrial scale worldwide. Karen Scrivener, Professor at the Polytechnic University EPFL in Lausanne, Switzerland, claims that more than 50 per cent of all man-made objects today consist in one or the other way of cement-bearing materials.[11] Mining, in contrast to cultivating, has a significant disadvantage: the resources do not regenerate themselves. All aggregates used in the concrete mix, sand and gravel, are forever lost and transformed without the possibility to recover them in their original form. In the

case of sand, the ruthless exploitation of this natural resource already shows dramatic consequences.

Sand is a product formed by a very long decomposition process of stone under the influence of natural elements. What is known and defined today as sand is the product of millions of years of enduring decomposition of stone materials in mountain areas, flushed through streams and rivers into our oceans. Yet the ever-increasing hunger of our building industry mines this natural resource with a breath-taking and unsustainable speed. According to John Milliman,[12] mankind today sources twice as much sand as is being produced naturally within the same period of time. According to the Swiss TV channel SFR,[13] the global market for sand is estimated at 15 billion tonnes per year, with a value of 70 billion US Dollars. The United Nations programme UNEP mentions 30 billion tonnes,[14] while the actual figures might be even higher. 50 per cent of that sand is being sourced before it even reaches the sea. Drastic forms of sand mining appear all around the globe. Beaches in North Africa are being depleted illegally (with one out of two beaches in Morocco affected), rivers are being dredged and ocean floors scraped. Landmasses collapse and islands erode. The consequences reach far beyond the actual mining area and leave devastating traces. Water levels fall in dredged rivers in India, Thailand, and Cambodia, destroying traditional settlements and modes of life. Dredging seafloors kills not only the base of entire ecosystems, but also affects distant sea regions with sediments suspended in ocean currents. Marine sand mining has devastating ramifications that will haunt gener-

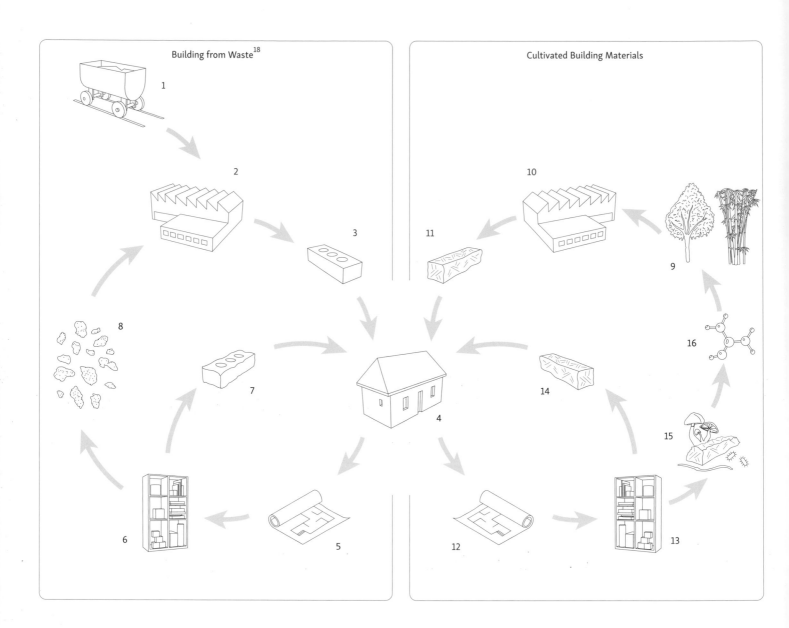

▲ In a circular building industry, all raw materials remain within the biological or the technical metabolism, turning from product to nutrient in a closed loop.

ations to come. Sand shortage leads to increasing illegal quarrying and trade in developing countries. "Sand wars"[15] and the "sand mafia" are commonplace terms in the media to describe the phenomenon.

Other materials as well are affected by our dominant mining mentality. There are minerals and metals in our earth's crust which are certainly even more scarce than sand. The German Ministry of Economy and Technology calculated back in 2005[16] that the global natural resources of lead, zinc, and tin, assuming a continued mining at current speed, would last less than 25 years. Even copper and iron become increasingly rare resources if we do not give up today's "take-make-throw"[17] mentality.

In recent decades, our society reacted to scarcity by becoming more efficient in order to sustain resources better. But why should we sustain a limitation? Should we not change towards an idea of abundant pertinence?

Metabolic thinking

The concept of cultivation implies the possibility of and wish for a "closed-loop" construction sector. For centuries, the "loose-ended" or linear model described above was the dominant rationale of the industrialized building material value chain. At first glance, the wood industry seems to be an exception here. However, the widespread use of chemicals, which started at the beginning of the 20th century, or the use of health-threatening glue matrices containing substances like formaldehyde,

▲ Edited by Dirk E. Hebel, Marta H. Wisniewska, and Felix Heisel, *Building from Waste* proposes a paradigm shift by acknowledging refuse as a resource for the building industry.

turned many of the supposedly biological wood products into hazardous or toxic waste. It is moreover not only the biological character of a building material which determines its metabolic behaviour, but also whether it can be returned after use to a state where it may become a nutrient for other cultivation processes. This type of circular or metabolic thinking started to become relevant in the 1970s, as we have described in our book *Building from Waste*.[18] Recyclability, biological compatibility, embodied or grey energy, a potential for repositioning, sorted reassembly, and other terms became criteria of value for an increasingly sensitized civilization. While the food industry easily adapted to such growing awareness and qualitative expectations, the building material industry is lagging behind. Phenomena such as the "sick building syndrome" only entered the discourse in the mid-1980s, when the US Environmental Protection Agency published a series of studies concerning indoor air qualities and their effect on the human body.[19] Yet still today, the majority of building materials is used without questioning their health impact. Michael Braungart, chemist and co-author of the manifesto "Cradle to Cradle",[20] reports that in his studies over the last decade alarming levels of toxic elements could be found in indoor spaces which were originally designed to bring comfort to people's lives. What is more, these materials are impossible to recover for a metabolic economy, and in the worst case their decay products infect our soils and bodies for generations to come.

Circular thinking sustains the flow of the materials without being altered or infected by other substances or products. Construction systems are re-quired to allow for the complete recovery of all materials during deconstruction, which may involve innovative joining systems or a complete avoidance of glues or unrecyclable composites. Walter R. Stahel, who is known to have coined the three R's – Reuse, Repair, and Remanufacture[21] – lately developed this idea further towards the 7 D's – Depolymerize, Dealloy, Delaminate, Devulcanize, Decoat, Deregulate, and Deconstruct –, reacting to a trend in the building material industry. Together with Werner Sobek of the Institute for Lightweight Structures and Conceptual Design (ILEK) at the University of Stuttgart, Germany, we promote the idea of "Building for Disassembly" to construct our homes in a way that allows any material to be recovered in a sorted and unpolluted way. Sobek and his team have introduced the Triple Zero: zero fossil energy, zero emissions and zero waste during dismantling or conversion.[22] Material, in this concept, is a borrowed good, which the client only borrows for a certain time and gives back to the market after use either as nutrition for a technical or biological cultivation process, or to be recycled or reused. This approach qualifies the initial investment cost for materials as secondary, as the resale value can be factored in. The medical machinery industry, for example, rents machines to hospitals while remaining the owner of the physical object. Once out of use, the manufacturer claims his resource to be returned, thereby closing the loop. Are such models conceivable for material use in the building industry?

Indeed, a new generation of cultivated building materials easily adapts to such a model: if a house can be grown, it could also be composted after its

use and become the source of a new cultivation process. We borrow material from a biological cycle and later feed it back into the cycle to be renewed. The price tag covers the use fee.

Cultivating construction

Some of the suggestions within this book operate with organic matter, which so far was seen as unwanted or labelled repulsive. For example, while the pharmaceutical industry uses bacteria with undisputed success, architecture and construction have not activated and exploited such capacities as of today. The same is true for mushroom mycelium. Another group of materials have been known for centuries as reliable and accepted resources for construction, yet they never advanced to the level of industrialized products – as if locked in a box labelled vernacular. Bamboo is the perfect example: high-rise scaffolding structures in Hong Kong, traditional buildings in South-East Asia and Japan, or highly individual signature architecture like a number of buildings by Simón Vélez, Jörg Stamm, Kengo Kuma or Vo Trong Nghia come to mind. For metabolic thinking, and viewed as a cultivated

building material, the potential of bamboo does not lie in applications of the raw material but in the extremely strong fibre that constitutes it. The approach of extracting and reconfiguring bamboo fibres is suitable to prompt the building industry to develop new industrialized production methods.

The cultivation of building materials requires a sustainable and ethical economic model. Coming back to the three principal ways of cultivating materials mentioned at the beginning of this introduction, especially the first of these options, i.e. soil-dependent cultivated materials, could have significant side effects when implemented within an uncontrolled and profit-driven agricultural model. A cautionary tale can be seen in developments in the palm oil industry, where natural forests are burned in order to bring more area under crops. A positive sign are the changing profiles of building experts. Multidisciplinary teams were engaged in the development of the building materials presented in this publication. Biologists, bioengineers, ecologists, chemists, and materials scientists cooperate with architects and civil engineers to create a

◄ The design of the Mirra office chair by Herman Miller implemented the cradle-to-cradle (C2C) protocol in order to eliminate waste and potentially harmful materials through planning. Designed for disassembly, the elements of the chair, at the end of its lifetime, can be fed back into a technical or biological cycle or reused for the same purposes.

▶ The construction of B10, a prototypical Triple Zero building by the office of Werner Sobek, took only 24 hours on site, due to modular prefabrication.

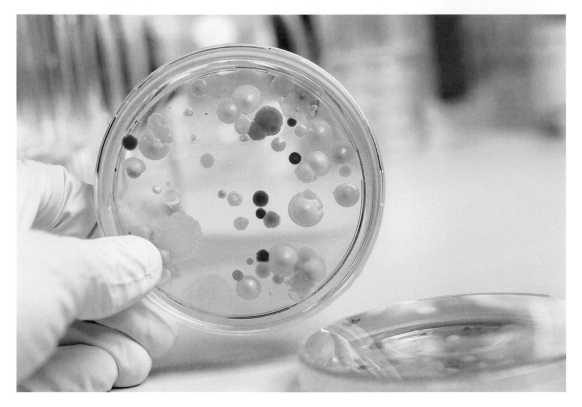

broader understanding of how to approach complex tasks. Supplemented by economists, this multidisciplinary work also holds the promise to sharpen our view towards alternative urban models, whereby production constitutes an integral part of a future urban society, calling for new types of spaces and infrastructures.

Cultivated Building Materials consists of three parts. The contributions of the first part present the approach of industrialized building with cultivated materials and its impact on sustainable construction and the building industry. Using as example the timber industry or the industrialization of rubber as well as the recent standardization of earth as a building material, this part showcases

ways in which a biological substance and its natural imperfection could be standardized, and the required circumstances and legal frameworks. It also addresses new and emerging economic models that could be associated with such a shift in production, connecting the discussion with issues related to the increasing urbanization worldwide.

In the second part, bamboo is used as a case in point. The research work on bamboo composites at the Future Cities Laboratory in Singapore, the Karlsruhe Institute of Technology KIT, Germany, and the Federal Institute of Technology ETH Zürich, Switzerland demonstrates the wide array of technological, social, political, and cultural questions connected to present and future

◄ **Mushroom mycelium offers a wide variety of properties for the cultivation of building materials.**

▶ **The Alternative Sand Lab of the Assistant Professorship Dirk E. Hebel at the Future Cities Laboratory in Singapore investigates possibilities to understand demolition waste products as resource for the production of biologically stabilized building materials.**

▾ **Cultivated bacteria can be a novel resource for the building industry.**

construction with renewable resources. The second part also discusses the potential of cultivated materials such as bamboo as mass-produced construction materials in the light of concurrent notions to use soil for food production rather than for cultivating construction materials. And it introduces a life cycle assessment of industrialized bamboo products with rather surprising outcomes.

The third part is conceived as a specific and tangible outlook into the future. It collects first uses and prototypical applications of newly developed cultivated building materials. Ranging from experimental developments to available products, building components and in some cases even structures, these case studies evoke thoughts, call for innovations, and invite the reader to connect the diverse input from scientific as well as empirical research to devise possible applications and outline a newly conceived architectural production process, asking a wider community to reply and respond. Here the book understands itself as a bridge spanning from scientific research to product development and building application. Neither a purely academic publication nor a product catalogue, Cultivated Building Materials describes the interdependencies, cooperations, and results of the joint efforts of academia and practice, incorporating a wide range of people from professionals to innovators in the field.

▲ The Hy-Fi tower was built as a temporary structure for the yearly Young Architects Programme at MoMA's satellite venue PS1 in Long Island City in 2014. It consisted of mycelium bricks, which were composted after disassembly. The project was a collaboration between the architecture firm The Living and the mycelium entrepreneurs Ecovative.

NOTES

1 Schwab, Klaus (2016), *The Fourth Industrial Revolution*. Geneva: World Economic Forum; remark from the authors: The four steps of the industrial revolutions are inspired by this publication.

2 Berz, Peter (2001), *08/15 – ein Standard des 20. Jahrhunderts*. Munich: Fink Verlag.

3 Ibid, p. 42.

4 Ibid, p. 49.

5 Canguilhem, Georges (1966/1991), *The Normal and the Pathological*. New York: Zone Books. Originally published in 1966 by Presses Universitaires de France in French language.

6 Ibid, p. 183.

7 Ibid, p. 200.

8 Ipid, p. 197.

9 Veblen, Thorstein (1919), *The Place of Science in Modern Civilization and Other Essays*. New York: B.W. Huebsch.

10 Cherenet Mamo, Zegeye (2015), "Designing the Informal: Spatial Design Strategies for the Emerging Urbanization around Water Bodies in Ethiopia" (dissertation), Hamburg: Hafen City University, p. 208; see also: Hebel, Dirk (2016), "Materializ-

ing Informality". In: Felix Heisel and Bisrat Kifle, *Lessons of Informality*. Berlin and Basel: Birkhäuser, p. 166–170.

11 Scrivener, Karen (2015), Lecture at ETH Zürich at the chair of Marc Angelil, Oerlikon, Switzerland.

12 Milliman, John D, and Syvitski, James P. M. (1992), "Geomorphic/tectonic Control of Sediment Discharge to the Ocean: The Importance of Small Mountainous Rivers", *The Journal of Geology*, September 1, 525–44.

13 ECO Spezial: Sand – ein Milliardengeschäft (2014). SFR TV, Documentary, Zürich, Switzerland.

14 Peduzzi, Pascal (2014), *Sand, Rarer than One Thinks*. UNEP: Global Environmental Alert Service.

15 Term from: Delestrac, Denis (2014), *Sand Wars*. Documentary.

16 Frondel, M., Grösche, P., Huchtemann, D., Oberheitmann, A., Peters, J., Angerer, G., Sartorius, C., Buchholz, P., Röhling, S., Wagner, M. (2005), *Trends der Angebots- und Nachfragesituation bei mineralischen Rohstoffen*, Bundesministerium für Wirtschaft und Technologie (BMWi) Forschungsprojekt Nr. 09/05, Rheinisch-Westfälisches Institut für Wirtschaftsforschung

(RWI Essen), Fraunhofer-Institut für System- und Innovationsforschung (ISI), Bundesanstalt für Geowissenschaften und Rohstoffe (BGR), Germany.

17 Leonard, Annie (2010), *The Story of Stuff: The Impact of Overconsumption on the Planet, Our Communities, and Our Health – And How We Can Make it Better*. New York: Free Press.

18 Hebel, D., Wisniewska, M. H., and Heisel, F. (2014), *Building from Waste – Recovered Materials in Architecture and Construction*. Berlin and Basel: Birkhäuser, p. 10ff.

19 https://www.epa.gov/indoor-air-quality-iaq, accessed September 10, 2016.

20 McDonough, William; Michael Braungart (2002), *Cradle to Cradle: Remaking the Way We Make Things*. New York: North Point Press.

21 Stahel, Walter R. (1982), "The Product-Life Factor", Mitchell Prize Winning Paper 1982. The Woodlands, Texas: Houston Area Research Center (HARC).

22 Heinlein, Frank (2015). *Residentials by Werner Sobek*. Stuttgart: av edition, p. 10ff.

▲ *Daring Growth*, the authors' contribution to the 15th International Architecture Exhibiton, Venice Biennale, offers a view into the laboratories of ETH Zürich, TU Delft and Mycoworks San Francisco, showing the cultivation and production processes of bamboo, bacteria, and mycelium building materials.

PRODUCTION BACK TO THE CITY

Manufacturing and the New Urban Economy

Dieter Läpple

In today's urban debate, the widespread hypothesis of a post-industrial city ultimately includes the disappearance of industry, manufacture, and craft-work from our towns as a consequence of the culturalization and digitalization of the economy.

The opposing view considers the shift from an industrial to a service economy in countries like Germany to be a *transformation* rather than a replacement or *substitution* of manufacturing by services, bringing with it new interconnections and interdependencies between the two sectors. An efficient manufacturing sector is regarded by many authors to be a prerequisite for the strong growth of business services. In this sense, the relationship between manufacturing and services is a complementary one and not a substitution.[1]

However, in the cities of the USA the structural change looked very different. In spite of early warnings against the myth and the dangers of a post-industrial economy,[2] the USA experienced a structural change that was indeed strongly oriented towards a substitution of industry by the service sector. As a result of this development, the income gap in the USA has become wider than it has ever been before, which has brought about an erosion of the middle and, especially, of the working classes. Nobel Prize winner Paul Krugman talks about the "social collapse" of the white working class.[3] Many political commentators draw attention to "declassed" white workers' class as potential voters for Donald Trump, who, in his election campaign, promised to bring back the old industries to the USA through a policy of protectionism.

Harvard Professors Pisano and Shih refer to what is arguably the most serious problem of such a post-industrial development: the increasing erosion of the innovative capacity of the economy.[4] The strategy of relocating more and more manufacturing capacities to Asian countries, while trying to retain the important research, development, and design competencies in the USA, did not work in the end. Over the course of time, Asian "contract manufacturers" have succeeded in taking control of increasingly larger parts of the value-added chain by building up their own research and development capacities, not least thanks to massive state subsidies. Today, the USA has not only lost the ability to manufacture high-tech products, it is also increasingly losing its competence and competitiveness in the development and design of complex industrial products and services.

It seems to be one of history's ironies that, having for a long time criticized Germany for delaying the structural change of their economy in favour of the tertiary sector, the USA is now widely discussing the possibilities of a re-industrialisation for which Germany is implicitly regarded as a role model, since its relevant industry has hitherto enjoyed relative international success.

Is a return of the production to the city a realistic option? This return could be facilitated and enabled by emerging changes in the global economy, changing consumer behaviour, and the need to decarbonize the economy.

Prospects for a productive city – possible ways out of the post-industrial deadlock

In spite of de-industrialization processes, most cities have retained a critical industrial base which, at least in successful cities, is integrated in so-called "service-manufacturing links": an interdependent, interactive system of knowledge-intensive industrial and service functions. The transformation of traditional industries based on mass production into a city-compatible network economy is certainly far from complete. Yet beyond the trans-formation and preservation of existing industrial production, new technologies also enable new factories to be established once again in the heart of the city. However, this requires a reinvention of industrial production. In a study on "smart manu-facturing",[5] Van Agtmael and Bakker formulate that "manufacturing will not so much 'return', then, as be reinvented".

One of the most spectacular possibilities is current-ly offered by the latest digitally based additive

▲ Additive manufacturing opens completely new possibilities for bringing production back to the location of consumption. The example shows elements of the "FutureCraft 3D" adidas shoe coming out of the 3D printer.

▼ The adidas Speedfactory produces custom-made, individual, yet indus-trially produced sneakers at the heart of urban locations.

manufacturing process: the industrial 3D printer, which is predicted to encourage a relocation of globalized production back to the centres of consumption. The company adidas has outsourced the production of its sport shoes to Asia for many years but now intends to relocate ('insource') parts of its production back to Germany by building a new "speedfactory", a combination of "Industry 4.0" and 3D printer technology. Strictly speaking, it is not a relocation of existing mass production but a reinvention of a production system that can produce small quantities on demand on a just-in-time basis at lower costs. In future, this type of customized production will go to wherever the consumer and his individual needs are located. Alongside such high-tech strategies, where "engineering skills replace craftsmanship", there is also evidence of a renaissance in craftsmanship among manufacturers (of shoes, for example), who tailor their production to a clientele keen to buy sustainably produced, durable goods.

The questions raised by these new developments in urban production, apart from how realistic these trends and projects are, mainly concern employment issues[6] and the contributions to an ecological restructuring of the city and a decarbonization of the economy. Under this perspective, it would not be wise to let ourselves be carried away solely by the "new economies". Our urban work environments comprise not only knowledge and culture-based areas, but also a wide variety of conventional industries and types of work and production, some of which are doggedly persistent, some are gradually eroding, and others are in a process of regeneration in hybrid form, ready to play a central role in

the creation and stabilisation of low-threshold employment opportunities and sustainable production concepts. An important step in this direction is the promotion of local economies by fostering district and neighbourhood businesses. Local economies comprise a broad spectrum of "close-to-home" small and micro enterprises ranging from artisanal manufacturing and repair workshops, retail outlets, and healthcare to gastronomy and other sectors of social, domestic, and business-based services. The range of these enterprises also includes traditional craft and migrant businesses as well as alternative firms and one-person entrepreneurs of the "new economy".

There is no single coherent vision for strengthening locally embedded economies. What is essential is an openness to all possibilities of local development strategies, the ability to detect and evaluate local potentials, and an intelligent networking, underpinned by the development of projects which encourage innovation, stimulate learning processes, increase local socio-economic variety, and open up options for action. However, the decisive factor is a clear political commitment of the city to provide the necessary resources for these future investments. Locally embedded economies should not be left to market forces alone or to traditional business development strategies. Successful support must focus on, and engage with, the social and cultural specificities of the various work environments embedded in the local economy. Simultaneously, crossover strategies will be of vital importance, interlocking diverse policy fields by linking urban development with social and economic strategies.

◄ The Weltgewerbehof at the International Building Exhibition (IBA) in Hamburg (2007–2013) activated former and derelict garages for urban production and local economy.

► One of the small- and medium-scaled urban production facilities at the IBA in Hamburg.

"Urban Manufacturing", "Green Industries", and the "Maker Movement" – contributions towards the "next urban economy"

American cities, which have been exponentially affected by the problems of a post-industrial economy, have also rediscovered the economic and social significance of manufacturing and material production. For several years there has been a wide-ranging discussion about the possibilities and prospects of "urban production" or "urban manufacturing". This new policy field is embedded in discussions about the development of the "next urban economy", which should no longer be driven by consumerism and debt but by production and innovation. Examples of successful "urban production" include urban fashion and clothing networks in New York, the biotech businesses in Boston, and the food production in Los Angeles. They comprise a mix of small and medium-sized businesses which often rely on skilled craftsmen and are based on specific customer requirements and local demand.

Ron Shiffman with his research group[7] and Saskia Sassen[8] have pointed out an important field of development in urban production: the role of urban manufacturers as "silent partners" in the creative industry. Here, the traditional relationship between material production and services is reversed: historically, production-based services developed according to the requirements of the manufacturing or processing industry; today, creative services are often the driving force. The cultural industry of the theatre or music theatre, for example, needs scenery and costumes. Television studios employ huge craft workshops, making high-precision equipment for HD television production. The product designers need qualified craftsmen to produce prototypes from which to realise their designs.

▲ Manuel Bornbaum and Florian Hofer cultivate oyster mushrooms in a rather unusual way – using as soil coffee grounds from the *Wiener Kaffeehaus* resources in the city centre of Vienna.

The ecological turn and attempts at constructing a post-fossil society also provide important stimuli and potential for the advance of urban manufacturing. In the USA, the "Urban Manufacturing Alliance"[9] is a national initiative that uses best practice models to make acquired know-how more accessible within an increasingly wide circle of cities and neighbourhood groups. "Manufacturing may never occupy the dominant position it once had in our economy, but a healthy manufacturing sector will provide high-quality employment opportunities in the 21st century." The future will look more to urban manufacture and less towards traditional large-scale industry: "... innovation and growth are more likely to come from small, urban manufacturing networks, whose locations and density enable them to respond rapidly to the changing needs of markets, whether local, regional, or global."[10]

Extremely interesting is the FabLab movement, a non-profit-oriented makerhood movement which uses 3D printer technology and ensures their know-how is freely available through open source initiatives on the net. At a time when everything seems to become dissolved in a virtual reality, the digital technology of the 3D printer (or "rapid prototyping") facilitates the return of material production to daily life. *Fabrication Laboratories* are networked mini-fabrication facilities offering computer-controlled modelling and production tools, essentially laser cutters, milling machines, 3D printers, etc. This mix of computer and mini-factory turns out complete products to individual designs and, in future, should enable a decentralization of production to consumer locations. The motto of this open source and makerhood movement is: "the neighbourhood is our factory".

Although it is difficult to assess to what extent such euphoric visions are realistic, they nevertheless show a new sensibility combined with an interest in material production and a passion for materiality and making, not only among intellectuals and an internet avant-garde but also among neighbourhood groups and young people. Combined with other changes, this could well result in a movement that will bring production back to the city.

I am convinced that there are good reasons to become seriously involved with new forms of material production. Modern industry and urban manufacturers offer a broad spectrum of skills and are not only city-compatible but also city-affine. It is worth thinking about new links and collaborations between services, industry, the creative economy, urban manufacturing, FabLabs, as well as local and immigrant economies.

Our "online society" still seems to devalue, or rather to disregard, the importance of the materiality of things. This myth must be vehemently opposed. Material production, also in industrial forms, remains an essential foundation for the city and for the diverse metabolic processes with nature. It is therefore essential to our future: developing a post-fossil economy in which the cultivated building materials – as portrayed in this publication – could form a vital part of urban production.

▲ The 3D Print Urban Cabin by DUS Architects in Amsterdam aims to showcase the potential of additive manufacturing for the future of urban areas. The prototype is 3D-printed from recyclable filament and represents a step towards the first fully printed house.

NOTES

1 Cf. Priddat, B.P., and West, K.-W (ed.) (2012), *Die Modernität der Industrie.* Marburg; Van Winden, W./ Van den Berg, L./ Carvalho, L./ Van Tuil, E. (2011), *Manufacturing in the new urban economy.* London and New York.

2 Cohen, Stephen S., and Zysman, John (1987), *Manufacturing Matters. The Myth of the Post-Industrial Economy,* New York.

3 Krugman, P. (2016), "Return of the Undeserving Poor": *New York Times,* March 15, 2016.

4 Pisano, Gary P., and Shih, Willy C. (2009), "Restoring American Competitiveness", published in: *Harvard Business Review,* July–August 2009 (Reprint R0907S), S. 1–14.

5 Van Agtmael, Antoine, and Bakker, Fred (2016), *The Smartest Places on Earth. Why Rustbelts Are the Emerging Hotspots of Global Innovation.* New York.

6 Not the subject of discussion here is the widespread question of how many jobs are destroyed by a digitalization of the economy and the use of robots. For this, we refer to the brilliant article by David H. Autor (2015), "Why are there still so many jobs? The history and future of workplace automation", *Journal of Economic Perspectives,* Vol. 29, No. 3, pp. 3–30.

7 Shiffman, Ron et al. (2001), "Making it in New York. The Manufacturing Land Use and Zoning Initiative", Pratt Institute Center for Community and Environmental Development, New York, May 31, 2001.

8 Sassen, Saskia (2006), "Urban Manufacturing: Economy, Space and Politics in Today's Cities", in: *DSSW-Materialien,* Berlin.

9 (http://urbanmfg.org/)

10 Byron, Joan, and Mistry, Nisha (2011), *The Federal Role in Supporting Urban Manufacturing,* New York, NY: Brookings Institute and Pratt Center for Community Development, 30.09.2012, http://www.brookings. edu/research/papers/2011/04/ urban-manufacturing-mistry-byron.

MASTICATING MACHINE.

▲ Masticating machine for the production of rubber. Illustration from Thomas Hancock, *Personal Narrative of the Origin and Progress of the Caoutchouc or India-Rubber Manufacture*, London, 1857.

in his lecture at the Darmstadt conference, "Man and Technology":

"With regard to the new materials now available to us, they are submissive to a previously unheard-of degree. (…) These materials do not offer us any strict, specific qualities to comply with but say: please, you are the master, I am the servant; I'll do exactly whatever you want. (…) We are faced with a sovereign control over materials as we have never known before and must therefore confront a vastness for which we feel ill prepared."[7]

This quotation perfectly expresses the anxiety of designers, who must now take more responsibility for the form since it is no longer dictated by the properties of the material. Schwippert has clearly recognized that the latest technical development offered by plastics, with no fixed structure or mandatory treatment process, calls into question the very concept of "proper use".

For many 20th-century thinkers, even the word rubber – caoutchuk means "weeping wood" in the language of the native Amazonians – evoked a sticky product similar to chewing gum. This stickiness has become a metaphor for the general dissolution of fixed values in an increasingly "plastic" world. We are compelled to think of Semper's metabolic theory and his description of the resilience of caoutchouc when we read Jean-Paul Sartre's words: "Just to perceive the viscous, a sticky, unstable substance without equilibrium, is to be haunted by a fear of metamorphosis. To

touch the viscous is to run the risk of dissolving into stickiness."[8]

It is testimony to Semper's openness to new discoveries that he gives top priority in his *opus magnum* to a material designers are afraid to have anything to do with even one hundred years later. Caoutchouc has a subversive nature which makes it very difficult to accommodate in a strict typology of materials. The basis of the Semper system is a matrix comprising the four manufacturing techniques and the associated material categories corresponding to the four elements of the home: "the *hearth* as the center-point; raised earth as a *terrace* surrounded by posts, the column-supported *roof*, and the mat enclosure as a *spatial termination or wall*."[9] These four elements are products of the four original arts of the ancient craftsmen: textile art, ceramic art, tectonics, and stereotomy.[10]

Semper first describes these categories in his book, *The Four Elements of Architecture: a Contribution of the Comparative Study of Architecture*, published in 1851, in which he presents the harmonious interaction of the four elements as the key to the aesthetic perfection of the Greek temple.[11] In his *Style*, published almost ten years later, the dividing lines between these four categories appear much more elastic. Imitation performs an important cultural function: every material can take on the role of another; stone can imitate wood and vice versa. According to Semper's *theory of material change (Stoffwechseltheorie)* the origins of all the forms and motives created by

man actually lie in the primary techniques and materials but they become distanced from these roots as soon as they come into contact with new materials and techniques. As an example, "timber-style", i.e. a style idiom corresponding to the properties of wood, will undergo several transformations and, "by means of a (…) change of material evolve into stone-style (…)", writes Semper.[12] The walls of a house, originally woven in fabric or twigs, are preserved in brick patterns or a painted surface. The forms of the timber roof trusses are transposed into stone or iron. Caoutchouc, the "ape of materials", however, throws this fine systematic approach into confusion. It can be kneaded like ceramics, cast like metal, woven like fabric or applied as a liquid like a varnish. Its properties "can be varied indeterminately".[13] Therefore, it does not fit properly into any of these categories. The ape, symbol of imitation, seems to have been elevated to the status of demiurge. It possesses the ability to metamorphose, as indeed do all materials but to a lesser degree.

"A stylist faced with a material of this kind will be at a loss of words!" – writes Semper, not so much by rubber's lack of form, but rather by its ability to adopt or change form. We are not talking here about projecting or simulating a "creative will". Quite the contrary: rubber is able to imitate any material and to find within itself the optimum form to fulfil its function: "it yields to strong pressure but always springs back to its normal density. On the other hand, it is more easily stretched and is more inclined to stay in this condition."[14]

Designers of the post-war period such as Frei Otto or Heinz Isler adopted Antoni Gaudí's stereostatic models, also working with rubber membranes. Sergio Musmeci, an engineer born in 1926 in Rome and a student of Nervi, experimented in the 1950s and 1960s with minimal surfaces. The form of his Basento Bridge in Potenza, Southern Italy (1967–1969), was calculated using rubber foil models.

When Gaudí, Frei Otto and Musmeci turned the hanging model upside down to view it as a support structure, the reference criteria were also transposed: rubber translated into concrete, tension into compression. The form may be similar but it is an inversion. The same interplay between similarity and difference can be seen when Gaudí uses casts and matrices (a word which, like "material", derives from "mother"), concave items which reinforce similarity and create forms which evolve almost organically.[15]

This sheds light on a new dimension in the heuristic-constructive process. Rather than shaping a structure as if on a hypothetical potter's wheel, the architect-artist develops techniques, processes, and working hypotheses which then organize the material processes.

The soul of elasticity never completely disappeared from the classical avant-garde. The rigidity of works by Gerrit Rietveld and the De Stijl group is rather the exception. Hannes Meyer includes natural rubber, synthetic rubber, and other elastic materials and industrial products in his manifesto,

◄ **Plastercasts of natural forms from Antoni Gaudí's workshop.**

► *Accumulation (Häufung)*, **image from László Moholy-Nagy,** *Von material zu architektur*, **Munich 1929, p. 48.**

building, while stone, wood, and ceramics, i.e. Semper's "original materials", do not feature here at all. In his Bauhaus book, *von material zu architektur* (American edition: *The New Vision: From Material to Architecture*), László Moholy-Nagy uses car tyres to demonstrate "accumulation", "a material state often difficult to define", whereby substances build up like lava flowing from a volcano in New Zealand or a mass of open umbrellas viewed from above.

But it is first and foremost the modern Italian designers who were inspired by the *"anima di gomma"* rather than by the ghost in the machine. *Rubber Soul* was the title of an exhibition curated by the Pirelli Foundation in 2011, recounting the history of Pirelli clothing and advertisements. Compare this with the photographic collage created by Austrian artist Raoul Hausmann. Entitled *Elasticum* (1920), what we actually see is a man-machine hybrid. At about the same time, the Italian futurist Fortunato Depero was producing images of the modern worlds of rubber, including *Diavoletti di caucciù a scatto* (*Little Caoutchouc Devils*, 1919).

The rubber company Pirelli linked elastic material with the idea of modern comfort early on: road surfaces, furniture upholstery, car tyres were applications already featured in the company's production programme in 1933. Italy was compelled to intensify the search for new materials when, in 1935, the League of Nations, under pressure from France, England, and the USA imposed an embargo in response to the Abyssinian war. Financial sanctions, an export ban on raw materials, and an import ban on Italian goods all proved ineffective, however, and did greater damage to the League of Nations itself. The fascist regime elevated material self-sufficiency to its policy agenda. Owing to the threatening shortage of resources, bio-fuels were already being sought and new plastics developed to replace traditional materials. In architecture, Giuseppe Pagano[16] asserted that there was no wish to return to a marble classicism, despite the prevailing spirit of *Italianità*.

In 1936, coinciding with the VI Triennale in Milan, architect Franco Albini published a book, *La Gommapiuma Pirelli*, to illustrate the use of rubber foam in the furniture of Piero Bottoni, Gio Ponti, and in his own creations.[17] In his opinion, rubber foam has revolutionized the form of new furniture and has at last fulfilled the need for a suitable material to satisfy modern taste.

When the Second World War ended, the major Italian companies were determined that they, too, should benefit from rapidly growing economic competition. Intensive research into self-sufficient solutions through close cooperation between the materials research, industry, and design sectors continued to bear fruit. Montecatini, the large Milan-based chemical firm, had been supporting research into, and development of, synthetic rubber products and plastics since the beginning of the 1940s, the era of self-reliance. In 1957, polypropylene was first produced on an industrial scale under brand names such as Nailon, Vinavil and Terital. In 1952, Marilyn Monroe's consent to be photographed for the Pirelli-calendar in a Latex bathing suit was a milestone in the company's history.

Exhibitions, including *Formless Furniture* at the Museum für Gestaltung Zürich (2009), have clearly shown how little designer furniture by Frank Gehry or Gaetano Pesce have to do with the concept of formlessness. But should we ever be in danger of weakening the link between form and material as a result of ongoing material transformation processes, we shall find the way back to Semper's theory. His example of natural rubber demonstrates that material does not necessarily dictate form but that there can be free interaction between the two. Such freedom is not limitless, however, since all

new materials and products are integrated in a pre-structured system, sufficiently flexible to promote rather than restrict technical advances. The carnival mood evoked by Semper celebrates the interaction between the traditional and the innovative, the familiar and the unfamiliar. An ape makes us laugh because we see both the similarities and differences between his movements and our own.

This text is a revised chapter from the forthcoming book (2017) of the author, *Metamorphism: Material Change in Architecture.* Basel: Birkhäuser.

NOTES

1 Semper, Gottfried (2004), *Style in the technical and tectonic arts, or, Practical aesthetics*, transl. Harry Francis Mallgrave and Michael Robinson. Los Angeles: Getty Research Institute, 180.

2 Ibid, 169.

3 Ibid, 182.

4 Ibid, 183.

5 Ibid, 181.

6 Ibid, 182.

7 Schwippert, Hans (1952), discussion contribution. In: *Darmstädter Gespräch. Mensch und Technik: Erzeugnis, Form, Gebrauch*. Darmstadt: Neue Darmstädter Verlagsanstalt, 85.

8 Sartre, Jean-Paul, *Being and Nothingness. An essay on phenomenological ontology,* quotation from Didi-Huberman, Georges (1999),

"Die Ordnung der Materialien: Formbarkeiten, Schwierigkeiten, Kontinuitäten". In: *Vorträge aus dem Warburg-Haus 3*, Berlin: De Gruyter, 15.

9 Semper, Gottfried, *Style*, 666.

10 Semper, Gottfried (1852), "Plan eines idealen Museums". In Semper (1966), *Wissenschaft, Industrie und Kunst und andere Schriften über Architektur, Kunsthandwerk und Kunstunterricht*. Mainz, Berlin: Kupferberg, 72–29.

11 Semper, Gottfried (1989), "The Four Elements of Architecture. A Contribution to the Comparative Study of Architecture". In: Semper, *The Four Elements of Architecture and Other Writings*, transl. Harry Francis Mallgrave and Wolfgang Herrmann. Cambridge: Cambridge University Press, 74–129.

12 Semper, *Style*, 760.

13 Ibid, 181.

14 Ibid.

15 See Didi-Huberman, Georges (1999), *Ähnlichkeit und Berührung. Archäologie, Anachronismus und Modernität des Abdrucks*. Cologne: DuMont, 31.

16 Irace, Fulvio (2014), "Architettura e autarchia: Gio Ponti. Palazzo Montecatini". In: Finessi, Beppe, Miglio, Cristina, Eds., *Il design italiano oltre le crisi – Italian design beyond the crisis*, exhibition catalogue, Milano: Triennale Design Museum, Mantova: Maurizio Corraini, 93.

17 Bosoni, Giampiero (2014), "Franco Albini e la Gommapiuma Pirelli. Per una storia dello schiuma di lattice di caucciù in Italia (1933–1951)", *AIS/Design Storia e Ricerche*, no. 4, November.

THE INDUSTRIALIZATION OF TIMBER – A CASE STUDY

Nikita Aigner and Philipp Müller

There are many examples of building materials whose large-scale production and easy availability is the result of close collaboration of technological advances, scientific research, and standardization. Metals, ceramics, and synthetic polymers are among them. Timber, by contrast, is a singular example of a natural material having undergone this process. It proves that high efficiency and mass production, without necessarily destroying the natural resource of the forests, can be achieved even with a raw material of high variability and broad scattering of material properties. Although forest-based industries realized this fact only relatively late, when competing metal and ceramic industries already had an enormous lead in market shares of the construction business, this imbalance was intercepted in time, inviting us to tell the story of timber's industrialization process. This process can serve as a role model for other natural materials that have recently attracted scientific and industrial interest.

Image and industry

In most Western countries, the public perception of wood as a building material is ambivalent – a fact probably best summarized by Wladyslaw Strykowski's[1] laconic statement that "people love wood, but hate chainsaw [sic.]". Wood is generally perceived as a beautiful and environmentally friendly material, while at the same time the logging of wood is publically disapproved, at least in its most notable form of clear-cutting.[2] Similar to the case of meat products, there seems to be a certain degree of intellectual detachment from the technological processes and methods required to provide our everyday wood products. The transformation of forest-based industries towards an enormous degree of mechanization and optimization seems to have passed largely unnoticed by the public. Today, in the European Union alone, around 80 million cubic metres of sawn softwood are produced per year,[3] and wood industries – not including furniture – are generating an annual turnover of over 120 billion Euros.[4] A modern sawmill can process stems with speeds of up to 180 metres per minute, producing an output well over 1 million cubic metres of sawn wood per year.

▲ Approx. 80 million cubic metres of sawn softwood are produced in the European Union per year today.

▼ Forest-based industries have acquired an enormous degree of mechanization and optimization.

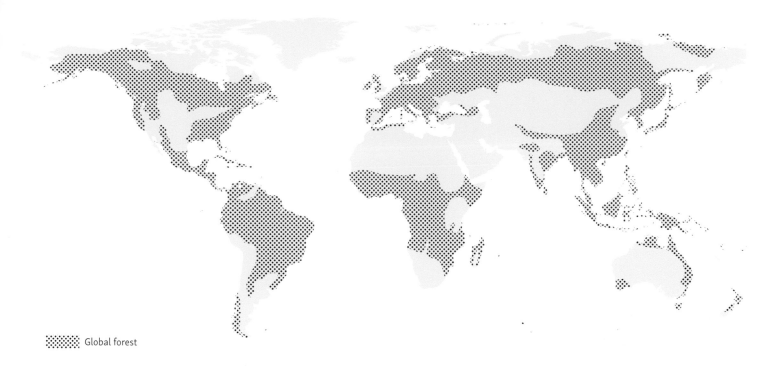

Global forest

Hand tools and industrial tools

As with most building materials, the development of wood processing to its current state is intimately linked to the development and availability of tools. From Prehistoric times throughout Antiquity, tool development experienced major advances with the discoveries of metallic materials – first gold, then bronze, and finally iron. While new tooling materials are still being developed today, mostly in the realm of ceramics, the discovery of steel was probably the most relevant material revolution in this regard. Starting with axes,[5] most hand tools have remained in principle largely unchanged since Classical Antiquity, but were subjected to millennia of refinement.[6] The results of this process are a myriad of variations within a relatively limited range of tools.[7]

The use of hand tools could never have resulted in timber becoming the industrialized building material that it is today. Hand tools are inherently slow and, indeed, imprecise compared to modern power tools. Some European joiners currently promote a renaissance of traditional craftsmanship that renounces the use of power tools. Although it is impressive to see the quality of objects created by these highly skilled artisans, a closer inspection renders their imperfections (even if they add to the aesthetic value of the piece) and the required manufacturing time unfit for modern large-scale production. Fuelled by the first industrial revolution, reaching its climax in the second half of the 19th century, major breakthroughs in sawmilling evolved through the introduction of new power sources.[8] Consecutive, incremental changes in sawing technology then transformed the steam-powered frame saw cutting by a few metres per minute into today's chipping lines.

While steel obviously played a crucial role in the process of cutting wood, it proved equally important in joining pieces back together. Even screws date back to about 400 A.D.,[9] yet did not experience widespread use until a standardized and hence mechanized fabrication became possible.[10]

▴ **Global timber habitat.**

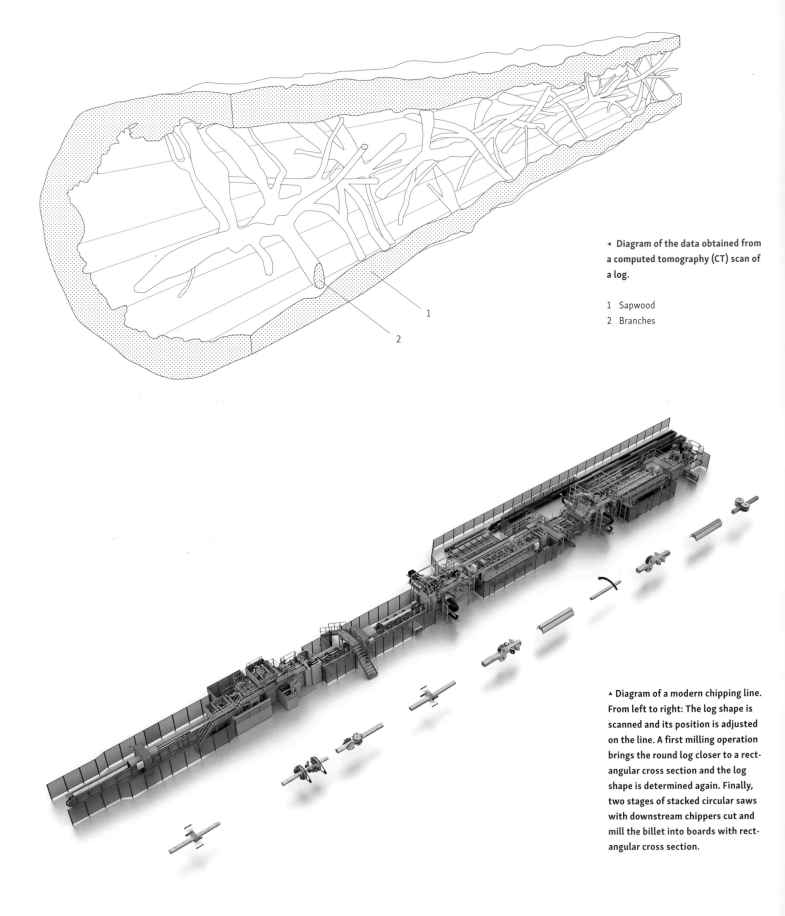

◂ Diagram of the data obtained from a computed tomography (CT) scan of a log.

1 Sapwood
2 Branches

▴ Diagram of a modern chipping line. From left to right: The log shape is scanned and its position is adjusted on the line. A first milling operation brings the round log closer to a rectangular cross section and the log shape is determined again. Finally, two stages of stacked circular saws with downstream chippers cut and mill the billet into boards with rectangular cross section.

Today, steel is still one of the dominant materials used for joining wood, and the family of metallic wood fasteners has constantly expanded to include, for example, plate and dowel systems.

The role of science and IT

The skyrocketing developments in electronics and information technology of the 20th century played a crucial role for the industrialization of timber, too. Important technologies included non-destructive testing (NDT) methods and digital image processing in the 1950s. Manual grading of boards became obsolete through the introduction of the first grading machine in 1969,[11] rendering grading not only faster but also more exact and reproducible by excluding human error. Due to high investment costs, these technologies are even today only used in large-scale sawmills, while smaller sawmills rely on manual grading techniques.

More recently, new developments were integrated at the front end of the sawmill. Computed tomography (CT) technology fascinated the community of wood scientists from the 1990s onwards,[12] allowing the investigation of the internal structure of a tree stem in great detail and thereby offering the enormous potential to optimize cutting patterns before opening the organically grown tree. It took approximately two decades until CT became fast enough to be integrated into the sawmilling process. Today it counts as the state of the art in the industry.[13]

While wood anatomy as part of botany has been studied since the 17th century, wood science is a relatively young discipline.[14] One of the first instances of a scientific exploration of wood properties is Hartig's work of 1794 on the combustion behaviour of different wood species,[15] driven by a pragmatic interest to find appropriate pricing for firewood as a fuel for ovens. Genuine scientific investigation of structure-property relations in wood did not start until 1910, when the United States Forest Products Laboratory in Madison, Wisconsin, was established.[16]

Wood science has always been closely associated with the corresponding industry. Major contributions of scientific research were the determination

▲ **Engraving of a historic steam-powered saw mill.**

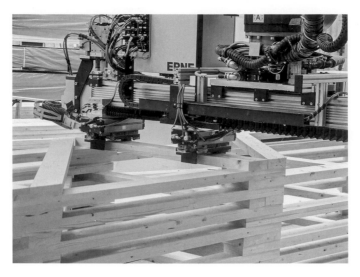

of mechanical properties and a detailed picture of the interactions of wood with water. These achievements laid the foundations for the implementation of NDT stress- and camera-based visual grading. Advances in wood chemistry allowed researchers in the polymer chemistry and adhesives industries to design optimized structural adhesives, and eventually paved the road for structural applications like glue-laminated timber (GluLam), plywood, and Oriented Strand Board (OSB). Structural adhesives likely gave wood the final push towards becoming a serious competitor for steel and concrete as a construction material.

Traditional wood glues suffer from the severe drawback of softening when in contact with water or heat, which renders them unusable for structural applications. Thermoset phenol-formaldehyde resins, developed around 1900, were far less sensitive to environmental conditions.[17] Due to its simplicity, low cost, and good adhesive performance, this material is still in use in many applications, including engineered wood products, but lately has been viewed critically due to health issues related to formaldehyde emissions. A multitude of other adhesive systems are tailored for specific applications and even specific wood species. Responding to the

◄ The structure is covered by a novel timber roof, which consists of 168 single trusses woven into a 2,308-square metre freeform roof design.

▶ A robot-based assembly process of 48,624 timber slats to individually formed trussed beams allows an efficient fabrication of the complex roof geometry.

▼ An integrated digital planning process connects design, structural analysis, the generation of fabrication data, as well as the manufacturing of the Arch_Tec_Lab, the new building of the Institute of Technology in Architecture (ITA) at ETH Zürich.

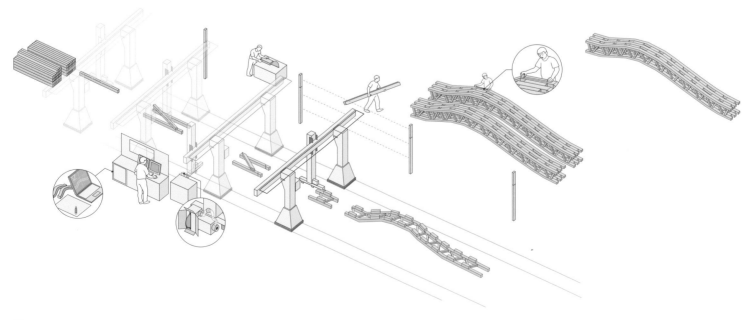

▴ The shell is constructed from the single module of a short and small-diameter beam, which allows for an optimal use of the wood as local resource.

issue of biological compatibility, this represents a field where new inventions are highly necessary.

Globalization, standardization, interdisciplinarity

None of today's industrialized wood products, however, emerged solely from science and technology. Another essential factor has been standardization, as it enables inter-organizational and international collaboration. Certainly the most sustaining contribution to a global reference system has been the metric system, whose implementation began in post-Revolutionary France to counteract the multitude and inconsistency of dimensions and units.[18] The American Society for Testing and Materials (ASTM) was established in 1898[19] and only a few years later, in 1904, a section devoted to wood was established, underlining this material's importance for industry. Today, ASTM lists almost 7,000 documents related to wood, while its European counterpart, the European Committee for Standardization (CEN), provides 440 standards. Including all national standards, there are more than 25,000 documents related to wood and wood products.

The current showcase product of industrialized timber is still GluLam, which was patented almost 150 years ago in 1872. Since that time, the introduction of new products seems to have been hindered by a number of factors. Being an organically grown material, wood is naturally unpredictable and inconsistent and therefore difficult to integrate into industrial processes that require constant quality and rely on standardization processes. In addition to this intrinsic material property, wood science's strong ties to the industry made it a very conservative discipline in which only relatively little incremental innovation has been possible. It is only on a recent trend that wood science opens towards more interdisciplinary scientific fields. Advances in computational methods, for example, allowed the recent development of methods to use naturally shaped logs in structural applications.[20] Current work on the chemical modification of wood, based on sophisticated analytical methods,[21] showed promising results for example in increasing the durability of wood products without the use of external protective coatings. Yet it will likely take decades until results of this kind find their way into industrial applications.

▲ The curved roof seamlessly integrates illumination and infrastructure.

NOTES

1 Strykowski, W. (2013), "Wood – a substitute or a raw material substituted for", *Annals of Warsaw University of Life Sciences – SGGW. Forestry and Wood Technology* 84.

2 Bliss, J. C. (2000), "Public perceptions of clearcutting", *Journal of Forestry* 98 (12): 4–9.

3 Anonymous (2016), http://www.eos-oes.eu/en/sawmill_industry_facts_figures.php, last accessed on 5 August 2016.

4 Anonymous (2016), http://de.statista.com/statistik/daten/studie/270059/umfrage/umsatz-der-holzverarbeitenden-industrie-in-der-eu/, last accessed on 5 August 2016.

5 Пушкарева, О. Б. and В. Г. Новоселов (2007), "История развития обработки древесины, начиная с древних времен до настоящего времени." Деревообработка: технологии, оборудование, менеджмент XXI века: 283–295.

6 Nedoluha, A. (1960), "Geschichte der Werkzeuge und Werkzeugmaschinen", *Blätter für Technikgeschichte*: 243–280.

7 Hasluck, P. N. (1907), "Carpentry and Jinery Illustrated", Cassell and Co.

8 Thunell, B. (1967), "History of Wood Sawmilling", *Wood Science and Technology* 1: 174–176.

9 Dickinson, H. W. (1941), "Origin and manufacture of wood screws", *Transactions of the Newcomen Society* 22(1), 79–89.

10 Bennett, M. (1988), "Development of the machine-made wood screw in England, 1760–1860", *Conservation News* 35, 22–25.

11 Fewell, A. R. (1982), "Machine stress grading of timber in the United Kingdom", *Holz als Roh- und Werkstoff* 40(12), 455–459.

12 Habermehl, A., and H.-W. Ridder (1992), "Computer-Tomographie am Baum. I: Erkennen der Baumfäule und Gerätekonzept", *Materialprüfung* 34.10: 325–329; Steppe, K., Cnudde, V., Girard, C., Lemeur, R., Cnudde, J. P., and Jacobs, P. (2004), "Use of X-ray computed microtomography for non-invasive determination of wood anatomical characteristics", *Journal of Structural Biology* 148(1), 11–21.

13 Berli, C. (2014), "Wertschöpfungskette Rundholz", Bachelor thesis, HAFL Zollikofen.

14 Kisser, J. G. (1967), "History of Wood Anatomy", *Wood Science and Technology* 1: 161–164.

15 Hartig, G. L. (1794), *Physikalische Versuche über das Verhältnis der Brennbarkeit der meisten deutschen Wald-Baum-Hölzer*. Marburg: Neue akademische Buchhandlung.

16 Kollmann, F. F. P. (1967), "General Remarks on the Development of Forest Products Research and the Creation of Wood Science", *Wood Science and Technology* 1: 168–169.

17 Baekeland, L. H. (1909), "The Synthesis, Constitution, and Uses of Bakelite", *Industrial & Engineering Chemistry* 1(3), 149–161.

18 Wenzlhuemer, R. (2010), "The History of Standardisation in Europe", European History Online (EGO), published by the Institute of European History (IEG), Mainz. URL: http://www.ieg-ego.eu/wenzlhuemerr-2010-en URN: urn:nbn:de:0159-20100921441, last accessed on 24 August 2016.

19 Thomas, J. A. (1998), "ASTM – A Century of Progress", ASTM, USA.

20 Frese, M. and Blaß, H. J. (2014), "Naturally Grown Round Wood – Ideas for an Engineering Design". In: Aicher, S., Reinhardt, H.-W., and Garrecht, H. (eds), *Materials and Joints in Timber Structures*. Dordrecht: Springer Netherlands, 77–88.

21 Keplinger, T., Cabane, E., Chanana, M., Hass, P., Merk, V., Gierlinger, N., and Burgert, I. (2015), "A versatile strategy for grafting polymers to wood cell walls", *Acta biomaterialia* 11, 256–263.

STANDARDIZATION OF A NATURAL RESOURCE

The Example of Industrial Earth Building Materials in Germany

Christof Ziegert and Jasmine Alia Blaschek

After decades of oblivion, the use of earth as a natural building material – both in renovation and new construction – has become increasingly common practice again in Germany. This regenerative resource – a product of the constant alteration and sedimentation of the Earth's crust – greatly varies in composition and quality, depending on the location of extraction and the original material of decomposition. This characteristic brings earth building materials in close proximity to organic materials, as the properties of biological substances cultivated in earth also vary according to the soil composition. This variation was for the longest time the reason why the material was implemented only by skilled craftsmen in unique projects, as the preparation and handling could not follow prescribed recipes as in industrial production. When new cement-based materials entered the mass market featuring a normed and standardized industrial quality, earth constructions lost more and more market shares and have survived only in niche areas of special applications. Then the new nationwide trend, mostly based on a more critical understanding of informed clients regarding health issues and resource management, called for the introduction of prefabricated, standardized industrial products made from earth materials. Up to that point in time, detailed technical guidelines were still missing in many areas and it was often not possible for planners to precisely formulate the technical specifications for earth materials and components. As a consequence, there was no valid basis for the construction authorities or authorized experts to assess compliance with requirements. This changed in 2013, when the publication of DIN standards (German Standards Institute) for earth blocks and earth mortar for masonry and plaster led to fundamental changes in the field. Since then it has been possible to designate desired properties such as strength and sorption classes within a qualified specification. Furthermore, the warranties given on standardized products represent a safety seal for planners and contractors. And in times of increasing competition, even the manufacturers benefit from an official declaration of quality. For this reason, the introduction of a DIN norm for this regenerative resource could function as a blueprint for all cultivated building materials, be they organic or not.

Earth

Energy-efficient and sustainable construction without an excessive deployment of building technology is enabled by diffusion-open and capillary-active building components from natural materials. In recent years, earth has advanced to become a high-quality building material for modern architecture. Its aesthetic qualities go along with beneficial effects on the indoor climate and general well-being. Environmental properties include an unparalleled low energy balance of many earth building materials due to their almost infinite capability to recycling and upcycling. The generally high level of acceptance of Germany's earth building practice is also the result of ongoing quality-oriented initiatives in the fields of training, planning, publicity, and the development of quality standards in recent decades.

Only specific earth mixtures are suitable for construction purposes. Unfortunately, naturally occurring earth mixtures with ideal relative proportions

Global earthen masonry

▲ Global occurrence of earthen
masonry structures.

▾ Earth test cubes in the offices of
ZRS Architekten Ingenieure in Berlin.

of binding and non-binding constituents are rare to find. Consequently, considering the growing demand in modern earth construction, industrial products are in high demand. Nevertheless, builders and architects may expect to be asked by their clients whether the soil on the site is suitable for yielding appropriate material. With some experience it is certainly possible to determine the suitability of a soil, using simple field and laboratory tests. But this master knowledge system does not provide a global reference system necessary for the establishment of a norm. The concept of norming this naturally-occurring material can only address products not bound to a specific building site. Ready-to-use earth mixes like mortars or plasters, or products like earth bricks, were therefore the target of a first regulation and standardization in Germany.

Product standards for earth building materials

During the post-war years, tens of thousands of housing units were built out of earth building materials in both parts of divided Germany. Extensive regulations and first standards were implemented at that time, which embodied the contemporary state of earth building regulations in Germany. In the 1980s, when earth building experienced its environmental renaissance, these regulations were obsolete. In 1992, the German Association for Building with Earth (Dachverband Lehm e.V.) was founded in Weimar by ambitious manufacturers, distributors, design professionals, artists, clients, and other professionals who worked with clay. In 1996, when Germany's regional building ministry of Mecklenburg-Vorpommern called for updated regulations, the *Dachverband Lehm* was mandated by

▲ **Earth plaster surface after test of resistance to abrasion.**

the German Institute for Building Technology to formulate new building authority-regulated principles for earth building materials. At this moment, the stakeholders in this process maintained that a revision of the old standards was impractical. The formulation of a new set of guidelines was even deemed unfeasible. Instead, a new technical guideline – the *Lehmbau-Regeln*[1] – was formulated to document the state of the art of earth building materials and components used in the field of renovation and new buildings.

The *Lehmbau-Regeln* are part of the *Sample List of Technical Building Regulations* of the German Institute for Building Technology as well as part of the *List of Technical Building Regulations*, which were introduced in almost all federal states of Germany. The application of the *Lehmbau-Regeln* has been restricted to residential buildings with a maximum of two storeys and a maximum of two units.

This means that for applications outside these limits approval must still be directly clarified with the building authorities in the individual case. A significant number of projects used this latter option.

It was only in 2013 that the first new product standards for earth building materials – authored by the DIN Working Committee *NA 005-06-08 AA Lehmbau* – were developed, adopted, and published. These are:
– DIN 18945:2013-08 Earth blocks – Terms and definitions, requirements, test methods
– DIN 18946:2013-08 Earth mortar for masonry – Terms and definitions, requirements, test methods
– DIN 18947:2013-08 Earth mortar for plasters – Terms and definitions, requirements, test methods.
The new DIN standards for earth building materials pursue the goal of ensuring stability and service-

▲ **Testing of earth according to regulation: detection of the natural lime content.**

43

ability of the three products addressed. They define requirements and features to an extent comparable with conventional building materials. In addition, environmental characteristics have been taken into account, which is exemplary for standards in the field of mineral building materials. Extra attention was given to softer ecological criteria, such as the establishment of a procedure to determine the CO_2 equivalent value or the water vapour sorption capacity that affects the indoor environment. The admissible natural radioactivity, a characteristic that is required to be declared for all mineral construction materials, was also implemented at a very low precautionary guidance level, in accordance with the European legislative process.[2]

The product standards for earth blocks, earth mortar for masonry, and earth mortar for plaster can now be incorporated into the parent application standards. For example, the current version of the plaster application standard *DIN 18550-2:2015-06 Design, preparation and application of external rendering and internal plastering – Part 2: Supplementary provisions for DIN EN 13914-2*

for internal plastering mentions earth plaster as one kind of plaster, next to lime, cement, and listed polymer-bound plasters. As a next step, the *Dachverband Lehm* is planning to develop further standards for prefabricated earth building materials, specifically for earth panels, to be submitted to the standards committee on earth building (Lehmbau) of the German Standards Institute.

In general, the existence of DIN standards is a source of pride within the construction industry. Earth has become considered a *normal* building material.[3] The *Lehmbau-Regeln* still apply for all on-site-produced earth building materials that are neither regulated by these new DIN-standards nor industrially produced, for example rammed earth. This arrangement supports cases in which earth is used as a readily available on-site building material in a subsistence economy context. Telling from the authors' personal expertise regarding damages in earth building, such on-site constructions appear typically to be of very high quality. The earth mixtures are usually prepared by experienced, skilled craftsmen who are familiar with the specific textures of earth building materials.

◄ Testing of earth according to regulation: figure-8 test (binding force test).

► Testing of earth plaster mortar following German norm DIN 18947: resistance to abrasion (test apparatus according to Gernot Minke).

Results

Recent outstanding examples of earthen architecture have succeeded in generating a profound change in perception regarding the use of earth in construction, for instance spectacular rammed earth projects in Switzerland, Austria, and Germany, or the clay plaster surfaces of the Kolumba Museum in Cologne, awarded with the German Architecture Prize in 2009. Such published flagship projects are not limited to countries with standardization regulations and their success is often due to the knowledge of individuals as well as the impact of positive publicity. Yet exemplary buildings can hardly generate the scope of accepted use which arises from rules and standards enabling the broad

application of an industrialized natural resource. In this respect, Germany's earth building practice has set a positive example also for other, so far unregulated materials, be they cultivated or regenerative. Despite initial resistance, even in the ranks of the protagonists, Germany's newly implemented earth building regulations are now valued as a quality assurance in the field. For planners and authorized experts, it is now possible to precisely formulate and examine the technical specifications for earth building materials and components. As a consequence, the acceptance and thus the market sales of prefabricated earth building materials have continuously increased and other materials could follow this role model.

◄ **Kolumba Museum in Cologne, designed by Peter Zumthor.**

► **Interior clay plastering of the Kolumba Museum.**

NOTES

1 Dachverband Lehm e.V. [DVL] (ed.) (2009), *Lehmbau Regeln*. 3rd. edition. Wiesbaden, Germany: Vieweg + Teubner.

2 Ziegert, C. (2014), "Natürliche Radioaktivität von Lehmbaustoffen – Ergebnisse neuer Untersuchungen und Qualitätssicherung", *Wohnung + Gesundheit* 6/14.

3 Additional Information / Norms and Technical Data Sheets:
DIN 18550-2:2015-06 – Design, preparation and application of external rendering and internal plastering – Part 2: Supplementary provisions for DIN EN 13914-2 for internal plastering,
TM01:2014-06 – Requirements on earth mortar for plasters as a component.

TM05:2011-06 – Quality supervision of earth mixtures as a source of prefabricated earth building products – guideline.
TM06:2015-06 – Earth mortar for thin-layered plasters – Definitions, requirements, test methods, declaration.

BAMBOO, A CULTIVATED BUILDING MATERIAL

INTRODUCTION

Dirk E. Hebel and Felix Heisel

"Bamboo is one of the oldest construction materials: the history of building with bamboo probably begins with the history of Man's development, and it is most likely that it has not yet come to an end."

Frei Otto[1]

When Frei Otto made the above statement in 1985, he foresaw a global revival of the grass labelled today as "the timber of the 21st century".[2] Universities invest millions in research on this plant's potentials. It is certainly no coincidence that as a Professor for Lightweight Structures, Otto showed

such a specific interest in bamboo as a resource for construction. Bamboo is extremely resistant to tensile as well as bending and buckling stresses, a property that makes it, in terms of lightness and strength, one of nature's most extreme creations. Additionally, vegetal rods (especially as tubes) are not only effective from a structural point of view but also very economical when considering transfer of load over distance.[3] Benefits such as bamboo's incredibly fast growth (depending on soil, climate, and species up to 92 centimetres per day[4]) and the ability to capture large amounts of carbon dioxide from the atmosphere (in managed plantations)[5] further contribute, in times of resource scarcity and

▲ Raw bamboo culms stored after harvesting.

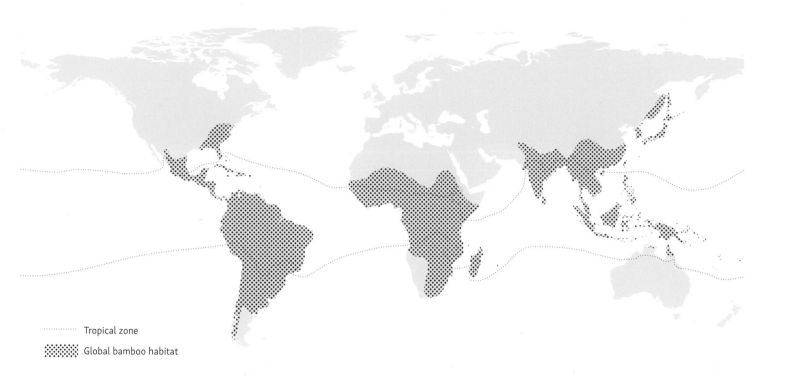

.... Tropical zone

▓▓▓▓▓ Global bamboo habitat

▲ **Global natural habitat of bamboo.**

climate change mitigation, to the current strong interest in this material. Above all, bamboo has always offered an affordable and accessible building material for the construction of homes.

Bamboos are grasses and belong to the family of *Poaceae*, which comprises a large number of economically important plants such as rice, corn, or sugar cane.[6] There exist about 1,200 to 1,400 bamboo species worldwide, distributed in about 90 genera; taxonomists do not always agree on their exact identification so that these numbers vary between sources. Geographically, bamboo is distributed in the tropical and subtropical belt between approximately 51° north and 47° south latitude[7] and can be found on all continents except Europe, which does not have any indigenous species.[8] Within this natural habitat, an estimated 37 million hectares of land are covered by bamboo forests: 3 million in Africa, 10 million in Latin America, and 24 million in Asia with 5 million in China, 11 million in India, and the majority of the remaining in South-East Asia.[9]

Bamboo can adapt easily to various soil and climate conditions. The plant consists of two main parts: the shoots and culms above ground and a dense root network or rhizome system below ground. There are two main types of rhizome systems, monopodial and sympodial. Monopodial rhizomes grow horizontally at an often surprising rate (hence their nickname "runners"). Their rhizome buds develop either upward (generating a culm) or horizontally (growing a new tract of the network). Consequently, monopodial bamboos generate an open clump with single, separate culms. These species are found in temperate regions and can be invasive due to their fast horizontal expansion. Sympodial rhizomes, on the other hand, are short and thick. Their culms stand close together in compact clumps, expanding evenly around its circumference. These species grow in tropical regions and are not invasive.[10]

Above ground, bamboo displays a surprising variety. While dwarf bamboo species reach only a few centimetres in height, medium-sized bamboo spe-

49

▾ **Symbodial and monopodial rhizome systems of bamboo. In the shoots, all nodes and internodes are already present and will elongate during the growth period.**

1 Monopodial rhizome network
2 Sympodial rhizome network

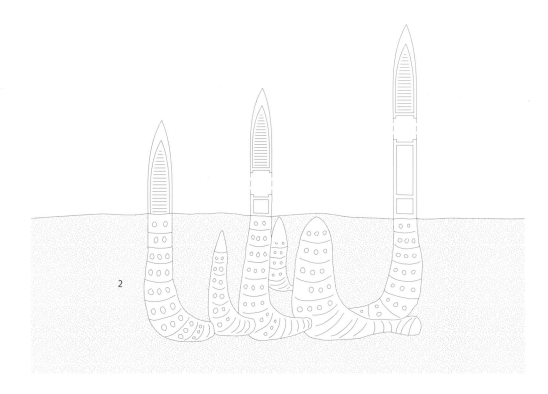

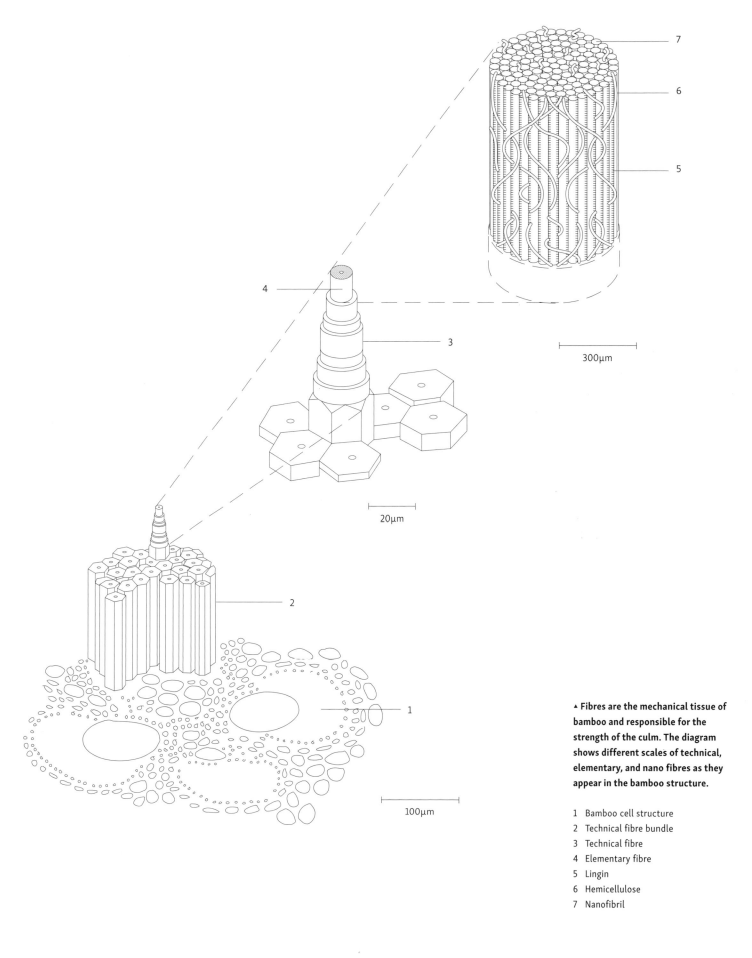

300μm

20μm

100μm

▲ Fibres are the mechanical tissue of
bamboo and responsible for the
strength of the culm. The diagram
shows different scales of technical,
elementary, and nano fibres as they
appear in the bamboo structure.

1 Bamboo cell structure
2 Technical fibre bundle
3 Technical fibre
4 Elementary fibre
5 Lingin
6 Hemicellulose
7 Nanofibril

cies may grow to several metres and giant bamboo species stand up to 30 metres with diameters of up to 30 centimetres. The culms are comprised of a main stem with filled joints (nodes) and hollow segments in between (internodes). Individual bamboo culms emerge from the ground and grow to their full height within a single season of three to four months, before small side branches extend from the nodes and leafing out occurs. Each shoot pushing out of the soil already contains all nodes, segments, and diaphragms which the fully grown cane will possess later. The segment closest to the ground increases in size first and the one at the top last. Once at full height, bamboo's cell structure begins to lignify until it is almost as hard as wood, yet more flexible and much lighter. Different from trees, which grow a new outer ring every year, the pulpy walls of bamboo culms slowly thicken towards the inside. Depending on the species, the initial shoots are considered a mature culm ready for harvesting after three to seven years.

The main components of bamboo culms are cellulose, hemicellulose, and lignin. The minor components are resins, tannins, waxes, and mineral salts.

The percentage of each component differs from species to species and depends on the conditions of bamboo growth and the age of the bamboo as well as the section of the culm.[11] Protected by a waxy outer skin, bamboo stems are hard and vigorous, allowing it to survive, and recover from, even severe calamities, catastrophes, and damage. Bamboo culms survived the nuclear bombings of Hiroshima and Nagasaki, and young bamboo shoots were the first signs of new plant life near the sites.[12]

One of the most thorough investigations into the various uses of bamboo was published by the Swiss salesman, collector, and author Hans Spörry.[13] Travelling to Japan on behalf of the Ziegler & Merian silk factory between 1890 and 1896, Spörry collected bamboo items and catalogued them for his friend Carl Schröter, who at that time was Professor of Botany at the Swiss Federal Institute of Technology ETH in Zurich. Spörry spent a lot of time and resources on gathering an unparalleled collection of about 1,500 objects made of, or referring to, bamboo. His 1903 book publication *Die Verwendung des Bambus in Japan* (The Use of

▴ **A bamboo shoot as it emerges out of ground.**

▲ Dense structure of natural bamboo forest.

Bamboo in Japan) lists 1,048 uses of bamboo for Japan only, a compilation that is still referred to by many scholars today. Another 498 ornamental variations were added in 1933 by Willard M. Porterfield, a scholar researching bamboo applications in China, bringing the total to 1,546 documented uses in Asia, respectively worldwide.[14]

Next to culinary, medical, spiritual, or decorative uses, construction is a very significant field of bamboo applications. According to the Food and Agriculture Organization of the United Nations (FAO),

an estimated one billion inhabitants worldwide live in traditional bamboo homes.[15]

By nature, only few bamboo species are suitable for construction. The selection depends on physical properties such as length and diameter of the culm, the distance between nodes, the strength and flexibility of the species, or the straightness of growth. Also questions of location and availability of the plant play an important role in the selection of the building material. These considerations result in a list of six main groups for architectural use: *Bam-*

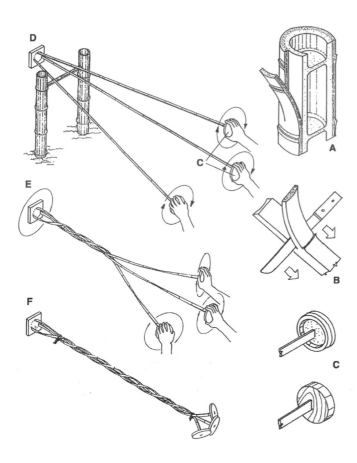

busa, *Dedrocalamus* and *Gigantochloa* in tropical Asia, *Chusquea* and *Guadua* in the tropical zones of the Americas and *Phyllostachys* in the temperate climates of China and Japan.[16]

Building with bamboo in the East is as old as the history of timber construction in the West – it simply represented the locally available resource for shelter. The first mention of bamboo as a building material in literature highlights two impressive Chinese constructions: as early as 221 BC, the palaces of the Qing Dynasty appear to have been built completely in bamboo.[17] Tension cables made from woven strips of only the outer layer of bamboo were used in shipbuilding and for the construction of suspension bridges, the most famous of which, Zhupu bridge, was reportedly built around 300 AD. This 261-metre-long bridge crosses the Minjiang River and connects the famous Erwang Temple and the well-known Dujiang Dam.[18] For its construc-

tion, builders used 500-metre-long bamboo cables, twisted from three or four towing cables with a diameter of 5 centimetres each.[19]

The West only learned about bamboo much later in history. In 1778, Carl von Linné, the Swedish botanist, physician, zoologist, and father of modern taxonomy, introduced the description by the term bamboo to science, based on the Indian word "Mambu" or "Bambu".[20] In 1789, a European botanist described the first bamboo genus (*Bambusa*) in customary Latin. By 1827, Japanese bamboo (*Phyllostachys nigra*) had been brought to Great Britain, while it took until 1882 for the first bamboo (*Phyllostachys aurea*) to arrive in the continental United States from China.[21] That same year, Thomas Edison opened the world's first light bulb factory using bamboo fibres as filament, creating probably the first industrial application of the material.

◄ Intertwined raw bamboo splits were already used as tensile cables in suspension cables and maritime applications in China around 300 AD. Columbian bamboo expert Oscar Hidalgo-Lopez displays their potential as concrete reinforcement in his book *Bamboo: The Gift of the Gods*.

► Bamboo is used as scaffolding systems in South-East Asia due to its strength, flexibility, and easy handling.

►► Traditional connection systems of bamboo joints (as shown by Gernot Minke).

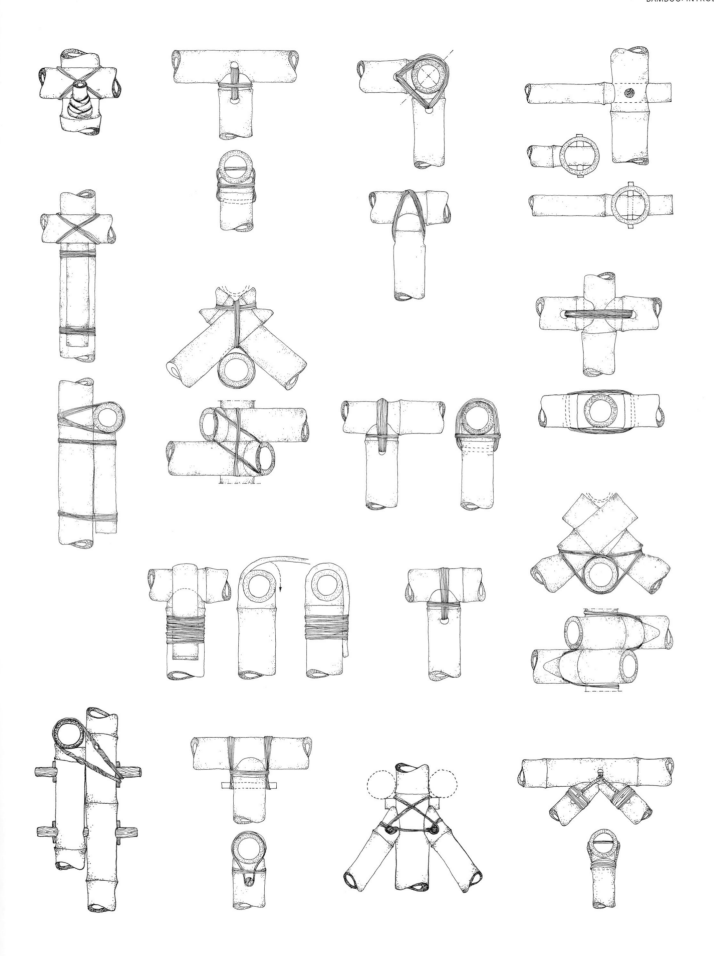

Since then, bamboo's strong, yet lightweight qualities have been explored by several inventions. The plant has been used to build boats and zeppelins, kites and early planes; in the Philippines, an entire plane was constructed exclusively from bamboo, while the Chinese commonly used it in their planes during the Second World War.[22] In the second half of the 20th century, the development of new and better industrialized materials, with carbon fibre and reinforced concrete at the forefront, shifted the interest of research away from bamboo as a resource for the future.

In recent years, the ecological and economic benefits of the plant reawakened the interest in its physiological properties. Universities outside the traditional bamboo territories, first and foremost the Federal Institute of Technology ETH Zürich in Switzerland and its Future Cities Laboratory in Singapore in 2012, but also the University of Cambridge in England, the Massachusetts Institute of Technology (MIT) in the USA, and the University of British Columbia in Canada as well as lately the Karlsruhe Institute of Technology (KIT) in Germany founded large-scale bamboo research institutes and laboratories. Their common agenda is to transfer the amazing properties of this organic resource, where each culm might differ from the others, into a durable, standardized, and always identical industrial product applicable within national and international building codes.

NOTES

1 Frei Otto (1985), "Preface". In: Gaß, Siegfried; Drüsedau, Heide; and Hennicke, Jürgen (eds.), *Bambus – Bamboo, Institute for Lightweight Structures (IL)*. Stuttgart: Karl Krämer Verlag, 10.

2 BBC (2012), "Booming Bamboo: The next Super-Material?".

3 Frei Otto (1985), "Preface". In: Gaß, Siegfried; Drüsedau, Heide; and Hennicke, Jürgen (eds.), *Bambus – Bamboo, Institute for Lightweight Structures (IL)*. Stuttgart: Karl Krämer Verlag, 10.

4 Farrelly, David (1984), *The Book of Bamboo: A Comprehensive Guide to This Remarkable Plant, Its Uses, and Its History*. San Francisco: Sierra Club Books.

5 Yiping, L., Yanxia, L., Buckingham, K., Henley, G., and Guomo, Z. (2010), "Bamboo and Climate Change Mitigation", *INBAR Technical Report No. 32*. Beijing: INBAR.

6 Oprins, Jan and Van Trier, Harry (2006), *Bamboo: A Material for Landscape and Garden Design*. Basel, Berlin, Boston: Birkhäuser, 13.

7 Minke, Gernot (2012), *Building with Bamboo: Design and Technology of a Sustainable Architecture*. Basel: Birkhäuser.

8 Hidalgo-Lopez, Oscar (2003), *Bamboo: The Gift of the Gods*. O. Hidalgo-López, 32.

9 Lobovikov, M., Paudel, S., Piazza, M., Ren, H. and Wu, J. (2007), *World Bamboo Resources: A Thematic Study Prepared in the Framework of the Global Forest Resource Assessment 2005*, No. 18, Non-Wood Forest Products. Rome: Food and Agriculture Organization of the United Nations, 12ff.

10 Ibid, 1.

11 Liese, Walter and Jackson, Bobby (1985), *Bamboos – Biology, silvics, properties, utilization*. Eschborn, Germany: Deutsche Gesellschaft für Technische Zusammenarbeit (GTZ).

12 Laidler, Keith (2003), "The Bear's Necessity", *The Guardian*.

13 Lüthi, Roland (2014), "Ein Schweizer Im Kimono". *ETHeritage*.

14 Liese, Walter and Michael Köhl (eds.) (2015), *Bamboo: The Plant and Its Uses*, Tropical Forestry. Cham, Switzerland: Springer International.

15 Lobovikov, M., Paudel, S., Piazza, M., Ren, H., and Wu, J. (2007), *World Bamboo Resources: A Thematic Study Prepared in the Framework of the Global Forest Resource Assessment 2005*, No. 18, Non-Wood Forest Products. Rome: Food and Agriculture Organization of the United Nations, 32.

16 Minke, Gernot (2012), *Building with Bamboo: Design and Technology of a Sustainable Architecture*. Basel: Birkhäuser, 9.

17 Yu, Xiaobing (2007), "Bamboo: Structure and Culture – Utilizing Bamboo in the Industrial Context with Reference to Its Structural and Cultural Dimension", doctoral thesis. Universität Duisburg-Essen, Germany.

18 Mikkolainen, Theri (2007), Chinese inventions: Suspension bridges. *gbtimes.com* Webpage, retrieved September 14, 2016, from http://gbtimes.com/world/chinese-inventions-suspension-bridges.

19 Dunkelberg, Klaus (1985), "Bamboo as a building material". In: Gaß, Siegfried; Drüsedau, Heide; and Hennicke, Jürgen (eds.), *Bambus – Bamboo, Institute for Lightweight Structures (IL)*. Stuttgart: Karl Krämer Verlag, 23.

20 Hidalgo Lopez, Oscar (2003), *Bamboo: The Gift of the Gods*. O. Hidalgo-López.

21 Farrelly, David (1984), *The Book of Bamboo: A Comprehensive Guide to This Remarkable Plant, Its Uses, and Its History*. San Francisco: Sierra Club Books, 5.

22 Adams, Cassandra (1998), "Bamboo Architecture and Construction with Oscar Hidalgo", Natural Building Colloquium Southwest, http://www.networkearth.org/naturalbuilding/bamboo.html, last accessed on 14 September 2016.

CONSTRUCTING WITH ENGINEERED BAMBOO

Dirk E. Hebel, Felix Heisel, Alireza Javadian,
Philipp Müller, Simon Lee, Nikita Aigner,
and Karsten Schlesier

Status quo: Raw bamboo in industrialized construction

Interest in bamboo as an industrialized construction material can be traced back to the year 1914, when PhD student H. K. Chow of the Massachusetts Institute of Technology (MIT) tested small-diameter bamboo and bamboo splits as reinforcement materials for concrete. Following this very first study, other researchers and research institutions around the world started experimenting with raw bamboo to replace steel reinforcement specifically for reinforced concrete beams and slabs. Among them was the University of Stuttgart in Germany, at that time a Technische Hochschule, where in 1935 Kramadiswar Datta and Otto Graf tried to find appropriate applications for the outstanding mechanical and technical properties of bamboo, unfortunately without any sustained success.[1]

In 1950, Howard Emmitt Glenn of the Clemson Agricultural College of South Carolina (now Clemson University) started to conduct more extensive research.[2] He and his team tested bamboo-reinforced concrete applications by erecting several full-scale buildings, utilizing his experience from work he had conducted in 1944 on bamboo-reinforced concrete beams. Using only small-diameter culms and bamboo splits, he demonstrated that the application was feasible; however, bamboo's elastic modulus, its susceptibility to insect and fungus attacks, the coefficient of thermal expansion, and a tendency to shrink and swell represented major challenges. Glenn applied different methods to minimize these defects, with varying success. Covering the bamboo components with coatings before placing them into the concrete mix, for example, reduced their exposure to moisture

▲ The SEC/FCL Advanced Fibre Composite Laboratory in Singapore conducts research on various species of bamboo to understand specific and varying characteristics of the grass.

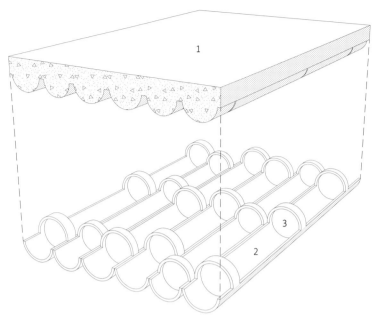

1 Concrete slab
2 Partially split bamboo as lost
   formwork
3 Interlocking diaphragms

and the alkaline environment of concrete. Eventually, however, the results showed that the solutions applied were rather expensive and the final product was unable to compete with steel reinforcement in economic terms. Besides, the long-term behaviour of bamboo reinforcement in concrete structures posed unsolvable problems for Glenn. As a natural material, bamboo absorbs water and moisture either from the moist environment of concrete during curing or through micro-cracks over longer periods of time. Water absorption leads to a progressive degradation of the bamboo culms and splits due to excessive swelling, resulting in a loss of bonding with the concrete. In the end, the structural members of Glenn's prototypes failed and the buildings collapsed.

In the following years, the United States military started to show growing interest in bamboo-reinforced concrete as an inexpensive and local solution for their increased construction activities in the tropics. In February 1966, Francis E. Brink and Paul J. Rush carried out a project for the US Naval Civil Engineering Laboratory at Port Hueneme in California in which they made suggestions for the design and construction with bamboo-reinforced

concrete as well as the selection and preparation of bamboo for the making of reinforced concrete.[3]

Based on these results, in 1970, Helmut Geymayer and Frank Cox from the US Army Engineer Waterways Experiment Section studied the application of a local bamboo species from Mississippi, *Arundinaria tecta*, for reinforcing concrete beams and slabs.[4] Concrete beams with a reinforcement ratio of three to four per cent were prepared, using bamboo splits that were either coated with an epoxy and polyester resin or pre-soaked for 72 hours before placing into the concrete matrix. The results showed that bamboo-reinforced concrete beams could develop about three to four times the maximum flexural strength of unreinforced beams of the same cross sections. As to cracking and failure, the results confirmed the conclusions by Glenn on the bonding behaviour of bamboo-reinforced concrete beams: due to different thermal expansion coefficients of bamboo and concrete and the ensuing thermal strains, cracking seemed unavoidable. This points at an important fact: while bamboo offers very high tensile strength, the low bonding strength between concrete and bamboo as a result of continuous swelling, shrinking, and differential

◄ Khosrow Ghavami used raw bamboo culms which were partially cut while leaving the nodes intact, to construct a lost formwork for concrete applications.

▶ Used in the construction of military hangars in Korea, raw and untreated bamboo splits de-bonded from the concrete matrix due to swelling and shrinking of the organic material and caused the structures to collapse (images published by Oscar Hidalgo-Lopez in *Bamboo: The Gift of the Gods* in 2003).

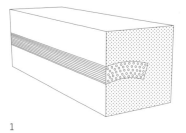
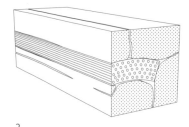
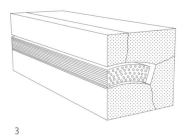

1

2

3

1 Natural bamboo placed in fresh concrete
2 Cracks caused by swelling of bamboo
3 Deponding due to bamboo shrinkage

thermal strains impedes the extensive application of bamboo as reinforcement for concrete elements.

In 1995, Khosrow Ghavami of the Pontifical Catholic University of Rio de Janeiro started a new series of mechanical tests on seven different types of bamboo. He hoped to determine which was the most appropriate species for use as reinforcement in newly developed lightweight concrete beams. The results demonstrated a remarkable superiority in terms of the ultimate applied load they could support compared to reinforcement with steel bars. However, the long-term behaviour of bamboo in concrete structures still remained problematic. Ghavami concluded that the different thermal expansion coefficients of bamboo and concrete would inevitably result in the de-bonding of the two materials.

Ten years later, Ghavami published a detailed study of bamboo reinforcement for a wide range of applications including concrete slabs, columns, and beams.[5] The bond behaviour of bamboo reinforcement to the concrete matrix was studied in this investigation through series of pull-out tests. Treated with an epoxy-resin bonding agent, the bamboo reinforcement showed an increased bonding strength that was 5.29 times higher than untreated bamboo reinforcement.[6] To create concrete slabs, Ghavami arranged treated, half-split bamboo culms next to each other before covering them with a layer of concrete, utilizing the bamboo as both tensile reinforcement and permanent (lost) formwork. To improve the mechanical interlocking between bamboo reinforcement and concrete, a composite interface was created between the two

materials by leaving the bamboo nodes intact when cutting the culms. Unfortunately, also these concrete slabs failed, mostly due to de-bonding and crushing of the culm sections and eventually due to concrete compression failure.

In the end, Ghavami's experiments have shown that replacing steel reinforcement in concrete applications is possible thanks to the superior mechanical properties of bamboo. But the problems with bonding and shear strength as well as durability of natural bamboo in the concrete matrix remained major barriers against a widespread application of bamboo as reinforcement in concrete applications.

Several other studies[7] have confirmed that natural bamboo could potentially replace steel in concrete applications, even though the ultimate load-bearing capacity of the concrete members reinforced with natural bamboo was less than 50 per cent of steel-reinforced concrete members. Yet these investigations did not elaborate on the main problem of using natural bamboo in concrete: (de)bonding.

In 2011, Masakazu Terai and Koichi Minami of Kindai University in Japan carried out series of pull-out tests as well as tests on bamboo-reinforced concrete slabs, beams, and columns.[8] A synthetic resin coating was applied on the surface of the reinforcements, which proved to enhance their bond strength. The reinforced concrete slabs showed signs of bond failure during the bending test, but the concrete beams and columns showed superior performance compared to non-reinforced members. Terai and Minami demonstrated that the

▲ **Effects of swelling and shrinking and the resulting de-bonding of untreated bamboo when used as a reinforcement system in concrete applications.**

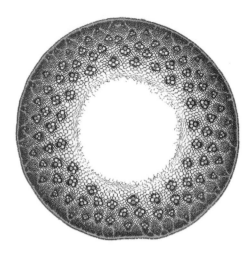

fracture behaviour of bamboo-reinforced concrete members can be evaluated by the existing formula for reinforced concrete members. But again, the results also indicated that the bonding strength between bamboo reinforcement and concrete matrix was only half of that of deformed steel reinforcement and concrete.

Makoto Yamaguchi, Kiyoshi Murakami, and Koji Takeda of Kumamoto University in Japan carried out a series of four-point bending tests of reinforced concrete beams with Moso bamboo reinforcement and stirrups.[9] The investigation showed that the load-bearing capacity of the bamboo-rein-

forced concrete beams could be predicted by section analysis based on the Bernoulli-Euler beam theory. However, this research did not investigate the effect of bamboo stirrups on shear failure and shear resistance capacity. Several further universities in Nigeria and India conducted similar research in recent years to evaluate the suitability of natural bamboo as a replacement for steel in concrete applications.[10] In all these investigations, bamboo proved to be a suitable replacement for steel reinforcement as far as its high tensile and flexural strength is concerned – yet did not measure up to steel in terms of bonding and durability.

▲ **Bamboo culm cross section showing the vascular bundle structure.**

▼ **Newly developed bamboo composite material at the SEC/FCL Advanced Fibre Composite Laboratory.**

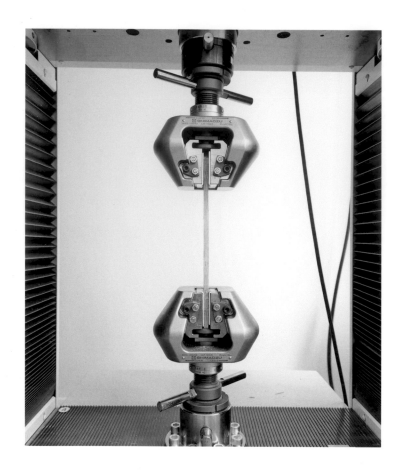

As a conclusion, the research conducted so far in the field of bamboo-reinforced structural concrete applications specified two major findings. Firstly, replacing steel with bamboo material as a reinforcement system is feasible. Secondly, so far, no solution is available to improve durability and control the swelling, shrinking, and thermal expansion of the material. In a way, these results are valid to the industrialized use of bamboo in the construction sector as a whole. Since the beginning of the 21st century, there has been a global effort to apply the developments from the timber industry to bamboo as a resource, fighting with the plants' very specific organic properties, as further described by Pablo van der Lugt and Felix Böck in their contributions (see pp. 72, 86).

Methods of engineering bamboo

Extensive research began in 2012 at the Future Cities Laboratory in Singapore and the Federal Institute of Technology ETH Zürich of Switzerland, exploring the development of a novel engineered bamboo material. Joined in 2013 by industrial partner Rehau and in 2014 by research partner Empa, the collaborative aim is to produce a high-strength, formable, water-resistant, non-swelling, non-corrosive, and durable biologically based composite material that takes advantage of bamboo's superior physical properties while mitigating its undesirable qualities. Departing from all former approaches, fibre bundles are extracted from the natural bamboo culm and treated to achieve the desired properties before binding them back together. The characteristics and the form of the resulting composite can be manipulated according to the desired product. Through such engineering, the biological bamboo material is being enabled for an industrial production process.

To this purpose the current research focuses on three areas: treatment of the fibre, appropriate adhesives, and a standardized production process.

The treatment of the fibre[11] is crucial to the success of the project. Only if the capacity of the fibre stays intact during extraction and composition, the

◄ Tension test of the newly developed bamboo composite material, performed in accordance to Eurocode and the standards of the American Society for Testing Materials, ASTM International.

► Pull-out test sample of the newly developed bamboo composite material.

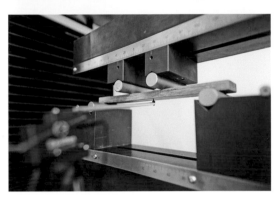

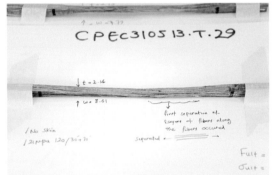

◄ Bone-shaped test sample of the newly developed bamboo composite material. The material is in accordance with the European standard, Eurocode, and the standards of the American Society for Testing Materials, ASTM International.

► Test samples before testing.

◄ Microscopic investigation of test samples and their behaviour at the SEC/FCL Advanced Fibre Composite Laboratory.

► Four-point bending test of the newly developed bamboo composite material, performed in accordance to Eurocode and the standards of the American Society for Testing Materials, ASTM International.

◄ All test samples of various test series at the SEC/FCL Advanced Fibre Composite Laboratory are documented, catalogued, and stored at the facility.

► Investigation of material behaviour.

naturally existing properties will be maintained. This is the most important difference from existing products in the market, which usually lose the inherent strength of the natural material as it is exposed to harmful treatment procedures. Tests on floorboards or kitchen tops of such kind, for example, show a dramatic decline of strength compared to the untreated bamboo material.

The second focus, the development of effective adhesives,[12] investigates the behaviour of and interplay between organic and inorganic materials.

The adhesive matrix controls factors such as water resistance, thermal expansion, and refractability.

The third aspect, the standardization of the production process, is crucial also for production in developing countries in order to guarantee safety factors and certify the product as a building material. As bamboo is still, in most countries, not considered to be an industrial, processed resource, standardization criteria and systems need to be developed so as to conduct research according to the strictest possible scientific terms.

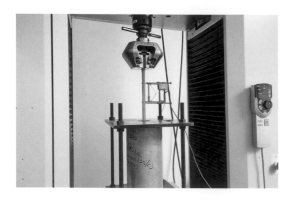

Engineered bamboo as a reinforcement system in concrete applications

Steel-reinforced concrete is the most common building material in the world. The developing countries use close to 90 per cent of the cement and 80 per cent of the steel consumed globally by the construction sector. At the same time, very few developing countries have the ability or resources to produce their own steel or cement, forcing them into an exploitative import relationship with the developed world. Out of 54 African countries, for instance, only two produce steel on a significant level above 4 megatons per year.[13] But steel is not irreplaceable. Bamboo is in principle available in those areas of the planet which are expected to have the highest share of construction activities in the decades to come, including Africa.

◄ Pull-out test of the newly developed bamboo composite material, performed in accordance to Eurocode and the standards of the American Society for Testing Materials, ASTM International.

► Four-point bending test of a bamboo composite reinforced concrete beam at the SEC/FCL Advanced Fibre Composite Laboratory.

▼ First applications of the newly developed bamboo composite material as a reinforcement system in concrete.

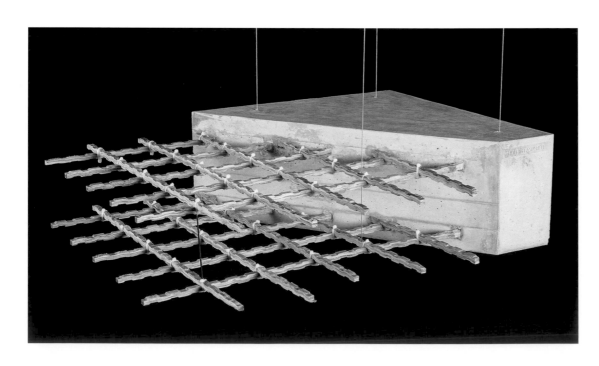

▸ Pull-out test samples at the SEC/ FCL Advanced Fibre Composite Laboratory. Here, the bonding between the new material and the concrete mix is tested.

▾ Testing scenario to measure the thermal expansion of the bamboo composite material in concrete.

Research at FCL Singapore and ETH Zürich investigates transforming natural bamboo into an easy-to-handle fibrous material, which is then fabricated into a high-tensile-strength composite using a hot-press method. Various species have been tested, all already used in construction and widely available. In each case, the raw material consists of an average fibre collection from the upper, middle, and lower sections of the bamboo culm, in nearly equal ratios. Also represented are the properties of all layers of a bamboo culm, acknowledging the varying properties of a naturally grown material in order to enable easy application of an industrialized process. Today, the Advanced Fibre Composite Laboratory at FCL Singapore is able to produce a bamboo composite which can compete with a low-grade steel reinforcement system in terms of tensile as well as bending strength.[14] The reinforcement system consists of longitudinal rebars and stirrups integrated into a concrete mix. In testing, these elements reach the same and even higher system properties in terms of bending capacity compared to a steel-reinforced system with identical reinforcement diameters.[15]

Moreover, extended pull-out test series proved that the newly developed bamboo composite material has highest bonding capacities with the concrete matrix, providing a solution to the described issues of durability in earlier investigations.[16] This is partially guaranteed by controlling the thermal expansion of bamboo through the adhesive matrix, as well as the composite's very low water absorption. The research so far confirms that by varying and controlling the process conditions (pressure, temperature, time) during

fabrication of the composite material, as well as enhancing the communication of the material towards other milieus (e.g. concrete's very high PH level), it is possible to tune the mechanical properties of a biological composite material into ranges that can in the future compete with tensile capacities of metallic as well as synthetic materials – at significantly lower cost. To reach this state, questions of standardization in the production process require further attention.

Multi-directional bamboo fibre-reinforced composites

In recent years, building industries around the globe have competed in developing lightweight materials with extreme tensile strength capacities. Materials of this kind find applications in aerospace industries, specialized building industries like mining and tunnel construction, and lately also in the automotive industry. All of these materials are composites. A symbiotic composite is typically

▲ Bamboo composite reinforcement system as tested at the SEC/FCL Advanced Fibre Composite Laboratory.

▾ Sample beams after testing.

▲ Vacuum moulding allows new application fields.

▶ Under vacuum, the adhesive is being pulled as thin and evenly as possible into almost any material configuration.

▾ Bamboo fibres could replace non-organic fibres in thin-shell applications.

created by mixing industrially produced synthetic inorganic fibres with an adhesive agent, resulting in a high-strength and at the same time lightweight material. However, conventional composite production is tedious and therefore extremely expensive. Raw materials such as glass or carbon fibres are hard to handle, represent a finite resource and usually require high amounts of energy in production, while at the same time they usually cannot be recycled and therefore end up as problematic waste after use. These are drawbacks that limit the access of such materials to high-end applications.

All renewable materials containing lignocellulosic fibres are promising candidates for creating a viable alternative to this situation: by developing bio-based composites that can be used for structural applications. Among these, bamboo fibres are a highly promising alternative to synthetic fibres due to their high mechanical qualities, low cost, cultivated abundance, and CO_2 storage capacity. So far, the major drawback for the application of bamboo in the composite industry has been the tedious and challenging fibre extraction process that reduces the mechanical qualities of bamboo fibres, making them weaker and therefore less attractive for the use in reinforced composites.[17]

The novel mechanical processing technique developed at FCL Singapore and ETH Zürich improves this fabrication process and its results. The developed bamboo fibre composites demonstrate that bamboo fibres can be used to create multidirectional fibre-reinforced materials. Similar to its synthetic counterpart, multidirectional fibre orientation within the composite can be achieved by

stacking up a number of bamboo fibres with specified angular orientations and adding a matrix system through vacuum infusion or a press. The result is a thin, yet strong and stiff multidirectional bamboo fibre composite that has properties comparable to glass fibre-reinforced plastics. This possibility to turn bamboo fibres into a multidirectional material is a crucial factor for the development of load-bearing applications: in the majority of applications, complex multidirectional loading states occur. By aligning the load-carrying bamboo fibres with the direction of the occurring loads, the load-carrying ability of the bamboo fibre composite can be increased and its mechanical properties optimized in terms of stiffness, strength, and weight.

The low cost of bamboo, its superiority in terms of biological availability over synthetic reinforcement fibres, and its good mechanical performance are key factors that lend multidirectional bamboo fibre-reinforced composites a huge potential. Bamboo as a cultivated alternative could replace commonly used structural materials in the automotive

sector, the building and construction sector, the wind energy sector (as a structural material for wind turbine blades[18]), or in the sporting goods sector. Furthermore, especially in developing regions in Africa, Asia, and South America where the scarcity of fossil and forest resources imposes the use of agricultural crops for the design and development of polymer composites, bamboo-based durable and strong composite materials will be of vital economic importance in the near future.

Bamboo-reinforced construction timber

Recent developments in timber construction, especially in the field of multi-storey residential structures, increasingly require innovative, high-performance, wood-based building materials and techniques. The bamboo composite materials as developed at the Advanced Fibre Composite Laboratory at FCL Singapore and ETH Zürich can be adopted to overcome shortcomings of wood, in particular its limited shear strength or brittle failure in tension, limitations which currently restrict the application of wood as a load-bearing material in building structures. There are three specific areas

▲ In the future, wind rotor blades could be manufactured out of natural fibres such as bamboo.

where bamboo composite reinforcements of timber structures may increase the performance of structural timber: bending reinforcement, shear reinforcement, and tension reinforcement perpendicular to the grain.[19]

Inlays of bamboo fibre composites into the joints of glued laminated timber represent a possibility to compensate for general shortcomings of timber such as a brittle failure in tension/bending and a high degree of variation in terms of mechanical strength. Inlays with a thickness of one to two millimetres can easily be placed in the joints during production to bridge imperfections such as branches or growth irregularities that are liable to cause failure, especially in the tensile zone. This bridging effect results in a lower degree of variation of the load-bearing capacity and thereby a higher degree of structural safety. As another result, material consumption can also be reduced. A particular feature is that the bamboo fibre composite inlays are able to shift the brittle failure in the bending-tension zone to a ductile failure in the bending-compression zone. In this way, overload can be monitored and failure can be predicted by the wrinkling of the fibres in the compression zone.

Bamboo fibre composite material is also of great interest for timber connection systems. Timber connections are commonly produced using steel plates and laterally loaded dowel-type fasteners, connections which are easy to design, produce, and assemble and yield a high load-bearing capacity. However, the heat transfer of metallic components leads to a rapid temperature increase in the connection, reducing the load-bearing capacity in case of fire and as a consequence the fire resistance of the entire structure. Furthermore, metallic components cannot be used in chemically harsh environments and where non-magnetic and/or non-conductive properties are required.[20] Bamboo fibre composites have the potential to replace steel as fastener in many applications.

The strengthening of timber with technical fibres by using finite resources such as oil (carbon fibres) or sand (glass fibres) requires large amounts of energy during production. This conventional combination threatens the major comparative advantage of wood as a fully natural, renewable, and recyclable building material. It will therefore be a crucial breakthrough to replace technical fibres with cultivated biological fibres for the reinforcement of timber structures. This is not only relevant for the strengthening of existing structures but above all can be applied in new timber-bamboo composite building elements. This combination makes use of the best properties of both materials, saving resources by introducing a high-strength lightweight material.

◄ Bamboo composite materials can be used as reinforcement elements in various applications.

► Test samples of the newly developed bamboo composite material.

Standardized products from cultivated building materials

Similar to the establishment of a material norm in earth construction, as described in the contribution by Christof Ziegert and Jasmine Alia Blaschek (see p. 40), research on bamboo composite materials needs to advance towards a standardization in order to develop a common determination base for designers and engineers. If this can be achieved, bamboo and other cultivated building materials have the potential to revolutionize the building industry.

The current homogeneous, world-spanning building industry no longer asks the most obvious questions: what materials are locally available? What is grown or cultivated? And how can I utilize these substances in the construction process? The quest for urban sustainability must be global in ambition, but it cannot be a matter of applying a universal set of rules and resources. Rather, sustainability requires a decentralized approach that acknowledges the global dimension and is yet, at the same time, sensitive to the social, cultural, aesthetic, economic, and ecological capacities of particular places to thrive and endure. Within such a setting, the most common plant growing in the developing regions of our planet will play an important role in economic and environmental terms if utilized as a cultivated and biologically compatible material by the building industry.

NOTES

1 Brink, F. E. and Rush, P. J. (1966), *Bamboo Reinforced Concrete Construction*. Defense Technical Information Center.

2 Glenn, H. E. (1950), *Bamboo Reinforcement in Portland Cement Concrete*. Engineering Experiment Station.

3 Yu, Xiaobing (2007), "Bamboo: Structure and Culture – Utilizing Bamboo in the Industrial Context with Reference to Its Structural and Cultural Dimension", doctoral thesis, Universität Duisburg-Essen, Germany.

4 Geymayer, H. G., and F. B. Cox (1970), *Bamboo reinforced concrete*. US Army Engineer Waterways Experiment Station.

5 Ghavami, K. (2005), "Bamboo as reinforcement in structural concrete elements", *Cement and Concrete Composites 27* (6), 637–649. doi: 10.1016/j.cemconcomp.2004.06.002

6 Ghavami, K. (1995), "Ultimate load behaviour of bamboo-reinforced lightweight concrete beams", *Cement and Concrete Composites 17* (4), 281–288.

7 Leelatanon, S., Srivaro, S., and Matan, N. (2010), "Compressive strength and ductility of short concrete columns reinforced by bamboo", *Sonklanakarin Journal of Science and Technology 32* (4), 419; Mahzuz, H., Ahmed, M., Ashrafuzzaman, M., Karim, R., and Ahmed, R. (2011), "Performance evaluation of bamboo with mortar and concrete", *Journal of Engineering and Technology Research 3* (12), 342–350;

Rahman, M., Rashid, M., Hossain, M., Hasan, M., and Hasan, M. (2011), "Performance evaluation of bamboo reinforced concrete beam", *International Journal of Engineering & Technology 11* (4), 142–146; Salau, M. A., Adegbite, I., and Ikponmwosa, E. E. (2012), "Characteristic strength of concrete column reinforced with bamboo strips", *Journal of Sustainable Development 5* (1), 133; Sethia, A. and Baradiya, V. (2014), "Experimental Investigation on Behavior of Bamboo Reinforced Concrete Member", *IJRET: International Journal of Research in Engineering and Technology 3* (2), 344–348.

8 Terai, M. and Minami, K. (2011), "Fracture behavior and mechanical properties of bamboo reinforced concrete members", *Procedia Engineering 10*, 2967–2972; Terai, M. and Minami, K. (2012), "Research and Development on Bamboo Reinforced Concrete Structure", Paper presented at the Proceedings of 15th World Conference on Earthquake Engineering, Lisbon.

9 Yamaguchi, M., Murakami, K., and Takeda, K. (2013), "Flexural performance of bamboo-reinforced-concrete beams using bamboo as main rebars and stirrups", Paper presented at the Third International Conference on Sustainable Construction Materials and Technologies.

10 Adewuyi, A. P., Otukoya, A. A., Olaniyi, O. A., and Olafusi, O. S. (2015), "Comparative Studies of Steel, Bamboo and Rattan as Reinforcing Bars in Concrete: Tensile and Flexural Characteristics", *Open Journal of Civil Engineering 5* (02), 228; Khan, I. (2014), "Bamboo Sticks as a Substitute of Steel Reinforcement in Slab", *International Journal of Engineering and Management Research 4* (2), 123–126; PaulinMary, B. B., and Tensing, Dr. D. (2013), "State of The Art Report on Bamboo Reinforcement", *IJERA: International Journal of Engineering Research and Applications 3*, 683–686; Pawar, S. (2014); "Bamboo in Construction Technology", *Research India Publications 4* (4), 347–352.

11 Hebel, D. E., Javadian, A., Heisel, F., Schlesier, K., Griebel, D., and Wielopolski, M. (2014), "Process-Controlled Optimization of the Tensile Strength of Bamboo Fiber Composites for Structural Applications", *Composites Part B: Engineering Journal 67*, 125–131.

12 Hebel, D. E., Heisel, F., Javadian, A., Wielopolski, M., and Schlesier, K. (2015), "Constructing Bamboo – Introducing an Alternative for the Construction Industry", *FCL Magazine Special Issue* "Constructing Alternatives – Research Projects 2012–2015 – Assistant Professorship Dirk E. Hebel (October 2015): 10–21.

13 Hebel, D. E., Heisel, F., Javadian, A., and Wielopolski, M. (2014), "Bamboo Reinforcement – A Carbon Alternative to Steel", paper presented at the World Sustainable Building Conference 2014, Barcelona.

14 Hebel, D. E., Heisel, F., Javadian, A., Lee, S., Müller, P., and Schlesier, K. (2016), "Engineering Bamboo – a Green Economic Alternative Part 01", *a+u* 16:05 (548), 162–167.

15 Hebel, D. E., Heisel, F., Javadian, A., Lee, S., Müller, P., and Schlesier, K. (2016), "Engineering Bamboo – a Technological Alternative Part 02", *a+u* 16:06 (549), 162–167.

16 Javadian, A., Hebel, D. E., Smith, I., and Wielopolski, M. (2016), "Bond-Behavior Study of Newly Developed Bamboo-Composite Reinforcement in Concrete", *Elsevier Construction and Building Materials* (122), 110–117.

17 Jawaid, M. and Abdul Khalil, H. P. S. (2011), "Cellulosic/synthetic fibre reinforced polymer hybrid composites: A review", *Carbohydrate Polymers 86*, 1–18.

18 Platts, M. J. (2014), "Strength, Fatigue Strength and Stiffness of High-Tech Bamboo/Epoxy Composites", *Agricultural Sciences 5*, 1281–1290.

19 Schober, K. U. et al. (2015), "FRP reinforcement of timber structures", *Construction and Building Materials 97*, 106–118.

20 Palma P. et al. (2016), "Dowelled Timber Connections with Internal Members of Densified Veneer Wood and Fibre-Reinforced Polymer Dowels", World Conference on Timber Engineering.

UPSCALING THE BAMBOO INDUSTRY

Felix Böck

Bamboo has the potential to be perceived as an innovative material of the future, rather than just another short-term trend. The modernization and upscaling of the production technology, or combinations of this outstanding grass with commercially known timber species for hybrid products, are currently changing the public perception and offer a potential for economic growth in many developing countries that have bamboo as a natural resource. The example of timber, described by Nikita Aigner and Philipp Müller in their contribution (see p. 32), demonstrates that the technical processing and the invention of specific machinery is an important and necessary step for the emergence of an industrialized product. While norms and standards can be employed to guarantee the material properties of large productions, new industrial processes with a new generation of machinery are needed to lay the foundations.

There has been a recent and increasing interest in the development of bamboo-based industrially manufactured goods, known as Structural Bamboo Products (SBP). Over the last decade, recognized institutions worldwide started to systematically review, investigate, and promote these engineered bamboo products.[1] In this process, the technologies and know-how from engineered wood products have been adapted to provide a varied range of materials that use bamboo for a growing list of structural, functional, and design applications.

Most of these bamboo products originate from Asia and are known as Strand Woven Bamboo (SWB), featuring densities over 1.0 kilogram per cubic centimetre. They are used in applications where otherwise plywood, Oriented Strand Board (OSB), Oriented Strand Lumber (OSL), and other glue-laminated composite wood products are employed[2] –

▴ Strand Woven Bamboo (SWB) products such as kitchen tops or flooring boards are entering the world market. These are mostly produced in facilities located in South-East Asia.

▸ Bamboo splits are dipped into an adhesive bath.

▾ Large-scale presses are in use to produce Strand Woven Bamboo (SWB) products. Due to heavy exposure to heat and high pressures, the fibres lose substantial mechanical properties.

mostly as project-based substitute products with a view to pricing, weight, and availability. Successful demonstration projects with various SBP applications include interior flooring, wall cladding and roofing systems in houses, small educational buildings such as schools and libraries, and even airports.[3]

There are many different ways to process raw bamboo into a wide range of shapes. Bamboo culms can be mechanically separated, chopped into sections, split into strips or slivers, crushed or scrimbered into fibre mats/bundles, stranded into defined strand sizes with specific lengths, widths, and thicknesses, flaked or chipped into pieces, and refined into fibres. In the development of the current bamboo industry, many processing technologies have been either newly developed or partly adapted and optimized from wood-based panel manufacturing. Recent innovative approaches focus on the unique characteristics of bamboo, like for example transforming the naturally round bamboo stems into engineered materials. This enhances the chance of developing and manufacturing bamboo products even in applications where bamboo has not been considered before.[4]

Upscaling production technologies

All engineered bamboo products currently available on the markets are produced in mostly labour-intensive, manual or half-automated processes. Driven mostly by outdated equipment that has been continually and incrementally (re-)adjusted over time in order to react to developments in material processing, glue or press technology, the global bamboo industry today is facing its technical limits.

◄ **Section of a Strand Woven Bamboo (SWB) product.**

1 Bamboo
2 Solid splits/slats/strips
3 Crushed bundles/mats
4 Strands
5 Fibres
6 Particles
7 Vertical Laminated Board
8 Horizontal Laminated Board
9 Woven Strand Board
10 Parallel Strand Lumber
11 Oriented Strand Board
12 Medium Density Fibre Board
13 Particle Board
►►

▼ **Table 1: Comparison of bamboo manufacturers based on their annual production volumes before and after production optimization.**

Location	Product	Capacity in m³/a	Optimization & additional investment	Capacity in m³/a
Factory A – Phase 1 (2012) China	SWB cold press	4,000 m³ Initial investment: US$ 3.0 million	– Drying prior to pressing – Optimal press parameters No additional investment	5,060 m³ (2014) Higher volume through moisture control
Factory A – Phase 2 (2014) China	SWB cold press	5,060 m³	– New pre-processing – Optimal drying – Optimal press parameters US$ 110.000 for new machines	11,200 m³ (2016) Higher volume through pre-processing
Factory B (2014) China	SWB hot press	2,680 m³ Initial investment: US$ 3.4 million	– Drying prior to pressing – Optimal press parameters US$ 30,000 (material handling and fans)	8,000 m³ (2016) Higher volume through moisture control and optimal press programme and material handling
Factory C (2016) South-East Asia	SBP continuous	50,000 – 100,000 m³ Initial investment: US$ 21.0 million	– New gluing technology – New pressing technology	Planning and commissioning process (2016)

A study undertaken by the laminated bamboo flooring and strand woven sector has shown that the majority of engineered bamboo manufacturers are using cold-press or discontinuous hot-press technologies with rather low production volumes, averaging 8,000 cubic metres per year.[5]

With an increasing demand of construction materials composed of renewable resources, the option of upscaling the bamboo industry to higher volumes and more industrialized processes is dependent upon new production technologies and optimized management of the material supply. Table 1 indicates opportunities in existing factories which have not yet been fully developed. The example of the participating factories focuses on two major areas of production optimization: firstly, the moisture balance with its impacts on material flow throughout the entire process and secondly, the pre-processing equipment.

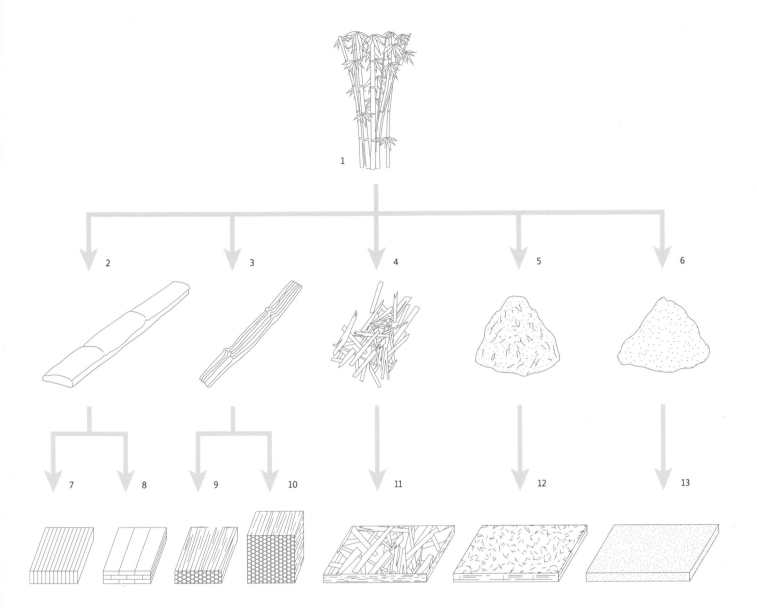

Compared to industrial production volumes in the wood industry, a more widespread use of SBP is currently limited by a lack of efficient technological synergies in manufacturing. The potential for reaching larger production volumes with products of a wider variety in terms of sizes and properties is being investigated. The example of Factory C in Table 1 describes a project with the aim to develop and activate a continuous production process for bamboo technologies by fabricating structural products with bamboo-based furnishes, focusing on strands, strips, and fibre mats. As discontinuous press sequences are the dominant production methodology today, this development obviously would increase speed and hence the output of the factory line to an estimated 50,000 to 100,000 cubic metres per year, a significant step compared to today's average 8,000 cubic metres per year.

Upscaling production processes

Currently marketed products, be they cutting boards or kitchen tops for an international market, are manufactured with discontinuous press technologies limited by specific dimensions in length of 2.40 to 3.60 metres and a density of 1,000 to 1,200 kilograms per cubic metre. In contrast, continuously produced bamboo-based composite panels/beams with densities closer to the relative density of the raw material (750–875 kilograms per cubic metre) are a new and innovative manufacturing approach, with applications in the construction sector, where high performances, longer spans,

▲ Industrial bamboo products use a wide variety of extracted raw materials, ranging from strips to fibres to particles.

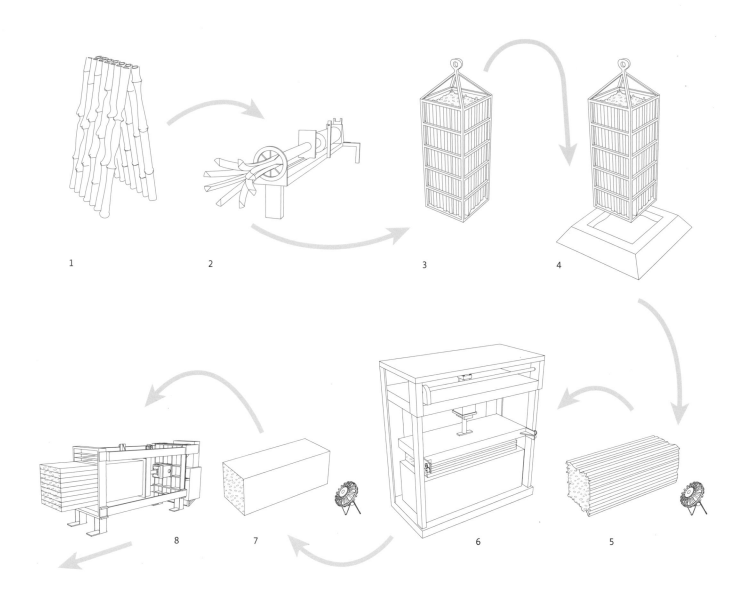

lighter materials but also cost competitiveness through higher production volumes are required. Similar production process technologies as known from Parallel Strand Lumber[6] (Parallam) and Oriented Strand Lumber (OSL) are the source of exploring adoptive synergies for the bamboo composite industry. This continuous process relies on a steady supply of raw materials which can be prepared in automated or half-automated processes, focusing on scaling up the pre-processing and pressing technologies of existing manufacturing processes.

Starting from the current batch production concept, a reduction in number of machines used for cutting, splitting, slicing, planning, and squeezing the raw material to its required shape and form is needed to achieve higher production volumes. Tests were undertaken using a variety of bamboo species, types, and dimensions, aiming to minimize tooling by maximizing throughput and crushing force. Aiming for high production outputs, cost savings in maintenance and tooling by achieving a higher material utilization, experiments were done to combine known processes that are bottlenecks and responsible for high expenses in tooling and material breakdowns. This knowledge is necessary for a continuous development of new production processes that are specific for the bamboo industry and can – next to the product itself – also create a new market for the machine building industry specialized on bamboo.

▾ Table 2: Material utilization factors in the industry compared to latest R&D outcomes. [7]

	Product	Utilization Rate (UR) in industry	Comments	UR after optimization/ industrialization
Non-structural & semi-structural products	SWB/WSB – scrimber bamboo products	~ 65% (based on Moso bamboo)	No utilization of bottom sections, utilization rate based on remaining culm	82–85%
	Laminated bamboo lumber and panels	~ 25%	Quality-sensitive process of material selection	25%
Non-structural & semi-structural products	Particle boards	> 80%	Residue-based production feasibility	82–85%
	SBP – scrimber-based structural beams and panels	N.a. (using continuous press technology)	Improved methodology of skin removal, treatment, gluing, and pressing	82–85%

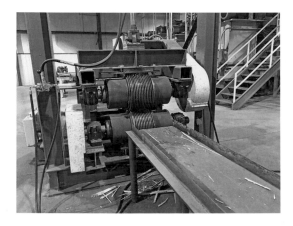

◂ Using automatic in-feed and a consistent rotation of heavy interlocking rollers with approx. 6,000 kilograms crushing force concentrated on a small split area, these grinders have a capacity of up to 5 tons bamboo per hour.

Upscaling material preparation technologies

The hardness of the bamboo skin, as well as its resistance due to its high silicate content, reduce the lifetime of tooling that is traditionally in use. Different approaches were investigated for the skin removal step prior to crushing that is needed to achieve the maximum bonding strength and product quality in durability and water absorption behaviour in the final product. A new process based on the use of a wire brush head has lately been introduced to remove the skin of the bamboo stem. It is a custom-fabricated roller that can be installed in sequence prior to the crushing step and is a method of pre-processing bamboo culms to improve automation, production volumes and, most importantly, reduce the residual waste.

Experiments have shown minimal losses in valuable fibre, while the pre-processing material loss has been reduced from 35 to 18 per cent (as described in Table 2). Tests were repeated using seven of the most commonly used bamboo species for engineered bamboo products, focusing on *Phyllostachys*, *Dendrocalamus*, and *Guadua* sub-species.

Another area of innovation is seen in the crushing of the bamboo material. Latest processing developments introduced machines with a capacity of 5 tons per hour, using automatic in-feed and a consistent rotation of heavy interlocking rollers with approx. 6,000 kilograms crushing force, concentrated on a small split area. These crushing rollers can vary in weight, size, and profile depending on the bamboo strip dimensions which are fed into the system.

Upscaling Material Supply Management

Recent developments in increasing the efficiency of production technology in the bamboo industry require a steady, non-fluctuating supply of bamboo culms in consistent quality as a raw material. Other than at centralized manufacturing areas, such as in the well-known area of Anji, China, individual projects face challenges in building up infrastructure for supply chain management systems. Other issues which are not addressed here are training and education as well as organizing cooperatives and farmers for selective harvesting methods.

When planning such a decentralized production facility for high-capacity manufacturing volumes (around 100,000 cubic metres per year), the following considerations need to be taken into account: (1) access to a manageable plantation area for local bamboo species with a minimum diameter of 10 centimetres at breast height; (2) an average clump density of 200 clumps per hectare, with eight culms per clump,[8] where (3) an estimated three culms can be harvested from each clump per year, yielding (4) an average green weight per culm of between 20 to 50 kilograms.[9]

With an approximate 10–12 per cent moisture content, each culm weighs about 6 kilograms, resulting in an estimated yield of 3.6 tons/hectare/year based on the average density of 650 kilograms/cubic metre. Considering the various utilization rates known for the use of different bamboo species, it is a reasonable assumption to make that approximately 50 per cent of the material is available in consistent quality for the production of engineered products. This implies that a 70,000-hectare plantation can yield approximate 200,000 cubic metres of raw material per year, a volume sufficient for the above-mentioned case study.

Cultivated industry

Critiques might mention that the cultivation of a new crop on an industrial scale of 70,000 hectares per factory can offset the ecological balance and take away land and nutritious soil from food crops in an economic competition. As Peter Edwards and Yvonne Edwards-Widmer show in their contribution (p. 80) on the ecological impact of bamboo, however, the grass holds indeed the possibility of an environmentally balanced resource even on an industrial scale. Not only is there already a significant amount of undermanaged bamboo forest that actually would benefit from a controlled harvesting, the plant is also known as a biological pioneer and grows on degenerated land where no other crops survive anymore. In this regard, land would benefit from bamboo's biomass in building up nutritious soil and by being protected from erosion through its dense root network.

Labelled with attributes such as organic, re-growing, strong, affordable, and available, bamboo thus has an incredible economic and ecological potential as a resource for future construction. Even so we can differentiate two different readings of this resource: firstly, the plant as it grows in culms – readily available for construction in local and unique (usually vernacular) building tasks; and secondly, the plant's fibres as a potential source of industrialized global building elements. The difference is fundamental and lies in the possibility of creating a standardized product through technological processing. No culm is ever the same (even when taken from the same species or family), varying in physical and aesthetic properties depending on age, soil, climate, and many other factors. However, when taken apart into fibre bundles and mixed with other specimens, the heterogeneous batch is then able to be certified within a range of specified properties – a precondition for standardization and mass-production as explained in the introduction of the editors.

Considering this, the upscaling of bamboo technologies is not only an economic question in terms of production volume, but equally important as an essential step in the creation of pre-fabricated Structural Bamboo Products for use in a global building industry. The case study in this chapter clearly shows the feasibility of the technology on an industrial scale. However, the research also highlights the still existing need for considerable technological developments in order to advance bamboo from the promising label of the "timber of the 21st century" to a certified, off-the-shelf product in construction.

NOTES

1 Frith, O. (2013), "Overview of Bamboo Sector & INBAR's New Strategy", UK Research Group Meeting, Bamboo construction for inclusive & green development, October.

2 Lugt, P. v. (2013), "The European Market for Advanced Bamboo Products – Opportunities for Africa?", African Bamboo Workshop, Addis Ababa, Ethiopia, March.

3 Zhang, Qisheng; Jiang, Shenxue; Tang, Yongyu (2001), "Industrial Utilization on Bamboo", *INBAR Technical Report No. 26.*

4 Boeck, F. (2014). "Bambus als Rohstoff – Potenziale prozesstechnischer Umsetzung", Holztechnologie 2014/05.

5 Boeck, F. (2014). "Green Gold of Africa – Can growing native bamboo in Ethiopia become a commercially viable business?". The Forestry Chronicle, 2014, 90(5), p.628–635.

6 Developed by Weyerhauser, Canada.

7 Liu, X., Smith, G.D., Jiang, 2., Bock, M.C.D, Boeck, F., Frith, O., Gatóo, A., Mulligan, H., Semple, K.E., Sharma, B., Ramage, M.H. (2016), "Nomenclature for engineered bamboo", Bioresources 11(1), p.1141–1161.

8 Tran, V. H. (2010), "Growth and quality of indigenous bamboo species in the mountainous regions of Northern Vietnam", doctoral thesis. University of Göttingen, Germany. Retrieved from https://www.deutsche-digitale-bibliothek.de/binary/

QESNNDOYZJMIXJE2ABLW3XWN-3PQF3VBV/full/1.pdf; Dang Thinh Trieu (2014), "Developing Guidelines for Sustainable Bamboo Forest Management in Thanh Hoa Province", *Vietnam Forests and Deltas Program Technical Report No. 19*; Thanh Hoa Provincial People Committee (2015).

9 Ly, P., Pillot, D., Lamballe, P., De Neergaard, A. (2012), "Evaluation of bamboo as an alternative cropping strategy in the northern central upland of Vietnam: Above-ground carbon fixing capacity, accumulation of soil organic carbon, and socio-economic aspects", *Agriculture, Ecosystems and Environment* 149: 80–90.

THE ECOLOGICAL IMPACT OF INDUSTRIALLY CULTIVATED BAMBOO

Peter Edwards and Yvonne Edwards-Widmer

As other contributors to this book have made clear, bamboo has the potential to become a globally important industrial building material, with obvious benefits for sustainability. But there is also a need for caution. In a crowded world, where is the land needed to support this new commercial crop? In particular, will growing bamboo take land that could be better used for producing food? How will bamboo plantations affect soil conditions and bio-diversity? We know from past experience that high demand for biological products can create prob-

lems. Large areas of woodland in Great Britain, for example, were converted to coniferous plantations to meet the demand for pulpwood during the 20th century. This often led to severe acidification of the soils and a drastic loss of biological diversity. Today, the rapidly increasing demand for biofuels is a matter of even greater concern, as it threatens not only global food supplies but also natural eco-systems, especially in the tropics. Recently, the Director General of the Food and Agriculture Orga-nization (FAO) drew attention to the increasing

▲ Bamboo forests in China are esti-mated to store from 169 to 259 tonnes of carbon per hectare, far more than the average 86 tonnes stored by timber forests around the world.

competition for land and water between food and non-food products, arguing that we will need to be much smarter in managing these resources in the future.[1]

Despite such pressures upon natural resources, especially land, there are also reasons to be optimistic. In the past, foresters considered it their main responsibility to produce timber and other wood products as cheaply as possible. Today, many see their job as managing forests to secure a much wider range of benefits, which include environmental services such as protecting watersheds, conserving soil and biodiversity, sequestering carbon, and providing places of scenic beauty for recreation and tourism. Economists have shown that if these services are properly valued, they often turn out to be worth more than the equivalent in timber that could be produced. This type of analysis is essential for developing smarter strategies to manage natural resources in the future.

The same approach is also needed for establishing bamboo plantations. Rather than valuing such plantations merely for the raw materials they produce, we need to ask whether they bring other benefits to the community and the environment.

Ecosystem services of bamboo
Many bamboos grow rapidly, even on poor soils. The initial growth of new shoots feeds on food reserves stored in underground rhizomes, so that the harvested biomass can potentially be replaced within a year. The continuous supply of shoots or culms which plantations can provide has been estimated at 10 to 30 tonnes per year depending

on species and climate.[2] Bamboos also produce a dense canopy of foliage and an ensuing abundant leaf litter. These features contribute importantly to the ecosystem.

Storing carbon. Plantations of bamboo are especially effective at creating carbon sinks. For example, bamboo forests in China are estimated to store 169 to 259 tonnes of carbon per hectare, far more than the average 86 tonnes that forests around the world store.[3] One of the main reasons for these high values is the large underground biomass of the rhizome system, which contributes to the organic content of the soil.

Stabilizing and improving soils. The carbon that accumulates in the soil beneath bamboo contributes to restoring degraded soil by way of retaining cations and by improving crumb structure and drainage properties. In addition, the tough and dense rhizome systems of some species, which can extend to as much as 150 kilometres per hectare, are extremely effective in stabilizing the soil, especially on steep slopes.[4] These effects, and also the fact that some bamboos can grow in very poor soils, make them ideal species for restoring degraded land, even on steep slopes which cannot be used productively for any other purpose.

Reducing floods. All plants take up water, which then evaporates through minute pores in the leaves in a process known as transpiration. In addition, plants intercept rainwater on their surfaces, and much of this also evaporates. Research has shown that rates of evapotranspiration, as the two processes as a whole are called, are particularly high in

vegetation dominated by bamboo. For example, stands of Moso bamboo (*Phyllostachys pubescens*) in Japan were found to evaporate far more water than the neighbouring forests, with the consequence that less water flowed into the rivers.[5] It was also shown that most of the water falling on the forest floor beneath bamboo soaked into the soil, so that only a small proportion ran off the surface. Both these effects mean that planting bamboo is an effective way to reduce the risk of flooding.

Cooling. Another potential benefit of the high rate of evaporation, especially in hot climates, is cooling the air. A dense stand of Moso bamboo, for example, can evaporate as much as 567 millimetres of water per year.[6] As an average rate of cooling, this evaporation is equivalent to 44 watt per square metre, though on some summer days the cooling effect could be more than twice as high. Clearly, bamboo can be very useful for cooling, though it may not be the best vegetation to have in places with water shortage.

Sustainable commercial production of bamboo

There are two kinds of places in particular where bamboo can be produced commercially without occupying farmland or destroying natural ecosystems.

Degraded land. According to the Global Environment Facility, nearly one third of the world's usable land has been degraded to such a degree that it suffers from reduced productivity.[7] In the tropics, at least, planting bamboo turns out to be one of the most efficient means of using degraded land productively and in ways that do not cause further damage but contribute to its restoration. Many bamboos improve soil quality and stability while at the same time providing construction materials and fuel. These are widely planted on degraded land throughout Africa and Asia. The International Network for Bamboo and Rattan (INBAR) has promoted several projects designed to produce bamboo commercially while protecting the soil. At an INBAR project site in China, soil erosion was reduced by 75 per cent within three years of planting bamboo and even before the canopy had fully closed.[8] For this reason, bamboos are especially useful for stabilizing steep slopes and riverbanks.

An inspiring example comes from Allahabad in India. The surface soil of an area had been stripped to make bricks, leaving behind land that was ruined for agriculture. In the 1990s, INBAR, in collaboration with the Indian NGO Utthan and the Indian Council on Forestry Research and Education, proposed a regreening programme based upon bamboo. Starting in two villages west of Allahabad in 1996, bamboo was planted to stabilize the soil on

▲ Bamboo planting is an effective reforestation method to prevent soil degradation also on poor soils.

Very high soil degradation

High soil degradation

Moderate soil degradation

Low soil degradation

the dykes around the excavated fields, and also in combination with other crops on the fields. Bamboo could even be planted successfully in the most degraded areas, where the soil had been burnt over many years during the brick-firing process.[9]

The successful project was extended to an area of some 4,000 hectares, covering the homes of 786,000 inhabitants in 96 villages. Today, bamboo forms part of several integrated cropping systems including leguminous trees, many types of fruit trees, vegetables, fish, fibres, medicinal plants, and livestock. Within these systems, where bamboo provides about 10 per cent of the total annual income of the farmers, the soil has improved greatly, containing more humus and nutrients and regaining its ability to retain water.[7]

Despite this spectacular success, restoring degraded land is a difficult and slow process that requires careful management. Soil carbon accumulates in the soil only gradually, as the roots and rhizomes decompose, and it may be necessary to burn the

bamboo to ensure that nutrients are returned to the soil in an available form. Not least, bamboo is notoriously difficult to get rid of once firmly established, so that changing to another land use may not be feasible.

Settlements and urban areas. In the past, bamboos throughout the tropics were planted close to settlements, providing not only construction materials, firewood, and charcoal but also shade and shelter. During recent decades, however, these plantings have largely disappeared, as urban dwellers have turned to other types of building material and other sources of energy. In many rapidly growing peri-urban areas, the overgrown and decaying remains of former bamboo plantations can still be seen, as a reminder of just how important they once were to the local economy.

Consequently, these growing urban settlements are now not only dependent upon fossil fuels and imported building materials but also suffer increasingly from problems such as flash flooding

▲ **Global areas of human-induced soil degradation.**

and warming due to the urban heat island effect. With an increasing realization that modern cities are unsustainable, we must find new ways of constructing and powering them using renewable sources of materials and energy, creating a strong case for the reintroduction of bamboo as a major component of sustainable urban ecosystems.

Bamboo also has other uses associated with settlements. One is for cleaning domestic wastewater by allowing it to flow through a filtration bed planted with suitable vegetation.[10] These root-zone treatment systems require little or no maintenance and are extremely effective in destroying pathogens. Such systems are in use in India, Australia, and many other countries, especially in rural and peri-urban areas that lack centralized water treatment services.

An important attribute of bamboo is that it can be harvested without destroying the plant. It can therefore be utilized as a source of timber and fuel while continually providing ecosystem services such as cooling, flood control, soil protection, biodiversity, and recreation. In addition, bamboo can be used as a feedstock for biomass gasification, providing clean energy in the form of hydrogen.

More research is needed to find the best ways of managing bamboo so that it can provide the fullest range of goods and services. While bamboo plantations should not be established at the expense of producing food, there are great opportunities for using bamboo to restore degraded land and improve the urban environment.

▲ Bamboo forests provide ecosystem services such as cooling, flood control, soil protection, sequestering carbon, biodiversity, and recreation.

NOTES

1 http://www.fao.org/news/story/en/item/275009/icode.

2 Liese, W. and Köhl, M. (2015), *Bamboo: the Plant and its Uses*. Cham, Switzerland: Springer International.

3 Song, X., Zhou, G., Jiang, H., Yu, S., Fu, J., Li, W., Wang, W., Ma, Z., and Peng, C. (2011), „Carbon sequestration by Chinese bamboo forests and their ecological benefits: assessment of potential, problems, and future challenges", *Environmental Reviews* 19: 418–428.

4 Liese, W. and Köhl, M. (2015), *Bamboo: the Plant and its Uses*. Cham, Switzerland: Springer International.

5 Komatsu, H., Onozawa, Y., Kume, T., Tsuruta, K., Kumagai, T., Shinohara, Y. and Otsuki, K. (2010), "Stand-scale transpiration estimates in a Moso bamboo forest: II. Comparison with coniferous forests", *Forest Ecology and Management* 260, 1295–1302.

6 Ibid.

7 https://www.thegef.org/gef/land_degradation.

8 https://hansfriederich.wordpress.com/2014/05/31/bamboo-for-land-restoration/.

9 International Development Research Centre (IDRC) (2014). "Bamboo's role in the environmental and socio-economic rehabilitation of villages devastated by soil mining for brick-making", *INBAR Working Paper no. 76*.

10 Steinfeld, C. and Del Porto, D. (2006), "Growing Away Wastewater", *Journal for Decentralized Wastewater Treatment Solutions* 2 (2). https://forester.net/ow_0603_toc.html.

ENGINEERED BAMBOO PRODUCTS – A SUSTAINABLE CHOICE?

Pablo van der Lugt

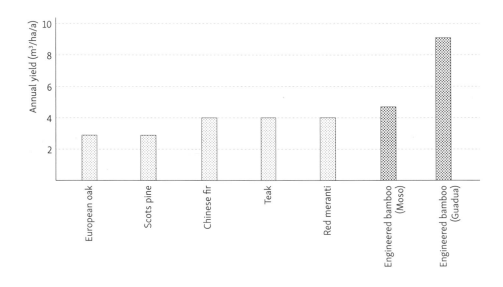

Given the increasing scarcity of resources, a transition from the traditional linear "make-take-waste" production scenario to a circular model is essential to be able to meet the needs of future generations. In the circular economy concept[1] products are designed in such a way that their components fit as nutrients in either the biological cycle – biodegradable after use, recycled, and regrown by nature itself – or the technical cycle of industrial non-renewable materials whose finite stocks need to be secured by high-level recycling.

Although the current focus of circular economy endeavours seems to be on incumbent high-tech industries of the techno-cycle, a shift towards the biological cycle or a bio-based economy seems more logical in the long term, due to its inherent renewable nature. With an increasing world population, and an increasing per capita consumption, human society will rely on high-yield cultivated building materials, such as bamboo, to meet future demand. Although sustainably grown wood will have an important role to play in this bio-based economy, in particular supplies of high-performance hardwood will be insufficient. Due to slow growth and long rotation cycles, hardwood forests will remain susceptible to deforestation.

As bamboo is a giant grass species, with many culms deriving from one mother plant and a fundamentally different way of growing and harvesting than trees (crop-like harvesting scheme based on annual thinning with high yields), it is less susceptible to clear-cutting, providing farmers with a steady annual income without a need of replanting. Furthermore, as Peter Edwards and Yvonne Edwards-Widmer describe in their contribution (see p. 80), bamboo is an excellent reforestation crop, even in areas of degraded land on eroded slopes, where farming is not feasible.[2] Clearly, bringing degraded land back to productivity is one of the key stepping stones to meet the increasing

(see p. 80)

▲ Annual yield of semi-finished material for various wood and bamboo species in cubic metres produced per hectare per year.

global demand for building materials, and bamboo can play a pivotal role herein.

In terms of versatility as well, bamboo fits perfectly into a bio-based economy. Multiple applications can be met with the same resource: not only can it be used to produce traditional (culm) or engineered building materials (beams, boards, and fibre products), but bamboo is also a perfect feedstock for the production of pulp (paper), fibres (textile), energy (charcoal), purifiers (activated carbon), and several utensils such as chopsticks, baskets, blinds, carpets, etc.

Engineered bamboo products

The industrialization of giant bamboo species such as *Phyllostachys pubescens* (known locally as Moso bamboo) commenced in the 1990s with the production of simple flooring boards made from laminated strips. Since then, the Chinese bamboo industry expanded quickly, adding many new prod-

ucts ranging from flooring, panels, veneer to solid beams. However, because of the limited durability and stability in terms of shrink and swell, applications were initially restricted to indoor areas. Recently, bamboo products suitable for outdoor use have hit the market, made from compressed bamboo strips that were first thermally modified (thermo-density treatment), offering a real alternative for tropical hardwood in outdoor applications. At the time of writing, the Chinese bamboo sector is the leading bamboo industry worldwide, employing 7.75 million people. With an export value of 1,207 million US Dollars, the country is responsible for 65 per cent of worldwide exports in bamboo products, and even for 97 per cent of the bamboo flooring market.[3]

Besides several other processing technologies under development and mimicking engineered wood products such as MDF, OSB, and several other composite products, four main technologies for

◄ **Flattened bamboo flooring boards.**

► **Laminated bamboo panels.**

◄ **Thermally modified bamboo decking.**

► **Strand Woven Bamboo (SWB) beams.**

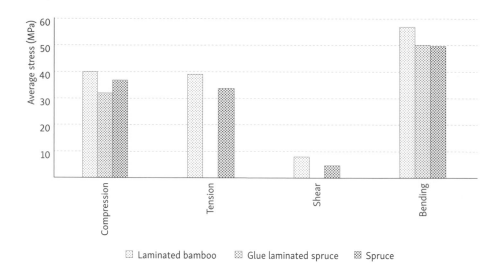

industrially processing bamboo to high-value-added engineered products can be distinguished in China at the moment, as Felix Böck describes in his contribution to this book (p. 72) in more detail:

1) *Flattened bamboo – 850 kg/m³*
   Flattening of longitudinally cut bamboo culms by vapour treatment, mainly used for the production of flooring.

2) *Laminated bamboo – 700 kg/m³*
   Lamination of strips to produce panels, beams, and flooring boards is the most commonly used technology to develop engineered industrial bamboo products.

3) *Cold-press compression moulding – 1,050–1,100 kg/m³*
   Compression moulding of rough bamboo strips with resin (cold-press) to extremely hard and dense boards and panels. These products are often also referred to as Strand Woven Bamboo (SWB) or bamboo scrimber.

4) *Hot-press compression moulding – 1,200 kg/m³*
   Same as 3) but based on hot pressing and with additional thermal treatment of input strips to increase durability and stability to enable outdoor use, mainly for cladding and decking.

Whereas most application areas for engineered bamboo were initially decorative, Western architects are increasingly exploring options to use engineered bamboo in (semi-)structural applications such as window frames and small structures such as carports. This applies particularly to laminated bamboo beams (finger-jointed on strip level, divided over the length of the beam for homogeneous composition) because of their relatively low weight compared to Strand Woven Bamboo. However, due to the industrial bamboo industry being at its very beginning, few harmonized structural tests have been executed to date. First results seem promising: studies executed by Cambridge University,[4] following EN 408 for a solar carport for German car manufacturer BMW, show that the applied laminated bamboo beams, impregnated to enable outdoor use, are comparable to and even exceed several important structural properties of typical wood species.

Life-cycle analysis of engineered bamboo products

The role that forests and wood or bamboo products play in the global carbon cycle is constantly gaining attention both positively (forest conservation, afforestation, increasing application in durable products – in particular in Europe and North America) and negatively (deforestation – in particular in tropical regions). It is estimated that around 47 per cent of the world's forest areas have been cleared

▲ Average stress of laminated bamboo compared to wood.[4]

or degraded to make way for crops, cattle, and infrastructure.[5] The greenhouse gas emissions resulting from deforestation account for 15 per cent of global anthropogenic greenhouse gas emissions. While illegal logging and clear-cutting of tropical forests for hardwood is part of the problem, the sustainable exploitation of tropical forests to produce certified timber (FSC/PEFC) is part of the solution, as it provides an economic incentive to

keep the forests intact rather than cutting them down for a onetime use. However, supply is often lower than demand, which is why additional hardwood alternatives are required. This is where engineered bamboo can play an important role. Giant bamboo species such as *Phyllostachys pubescens* are increasingly perceived as tropical hardwood alternatives because of rapid growth, good properties, and wide applicability. However, in a compari-

▲ The solar carport for BMW Group in South Africa is constructed with laminated bamboo beams, impregnated to enable outdoor use.

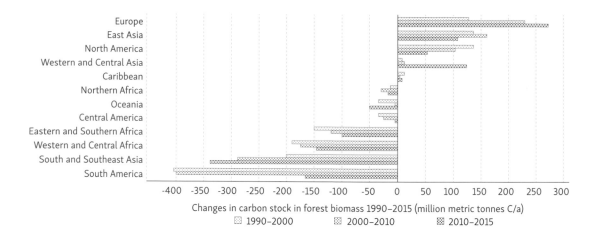

Changes in carbon stock in forest biomass 1990–2015 (million metric tonnes C/a)

▦ 1990–2000   ▦ 2000–2010   ▦ 2010–2015

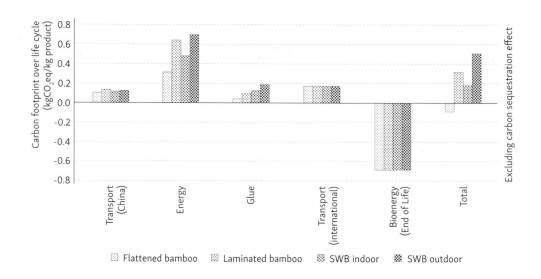

▦ Flattened bamboo   ▦ Laminated bamboo   ▦ SWB indoor   ▦ SWB outdoor

son with European wood, the relatively long production process and transport distance could disturb the benign environmental profile of bamboo.

In the following paragraphs, the greenhouse gas balance (carbon footprint) for the four main production technologies as mentioned above are analyzed following a cradle-to-grave scenario, i.e. including the full life cycle of the respective engineered bamboo materials:[6]

Production. To determine greenhouse gas emissions during production, all steps in the chain need to be monitored and calculated, from sourcing the

materials at the plantations in China, to processing, treating, and pressing in the manufacturing plant, to final packing and shipping towards their final location (in this case the Netherlands).[7] This cradle-to-gate analysis was based on best-practice production figures from the company MOSO International BV.

End-of-Life credit. Biogenic CO_2 is captured in wood or bamboo during the growth of a tree or culm. This storage remains intact as long as the resulting product exists. When discarded, carbon captured in the plant while growing is being recycled back into the atmosphere during the End-of-Life phase (e.g. composting). Consequently, wood

▲ Trends in carbon stocks in global forests from 1990–2015.

▼ Carbon footprint over life cycle (kgCO$_2$eq/kg product) for various industrial bamboo products based on different production technologies.

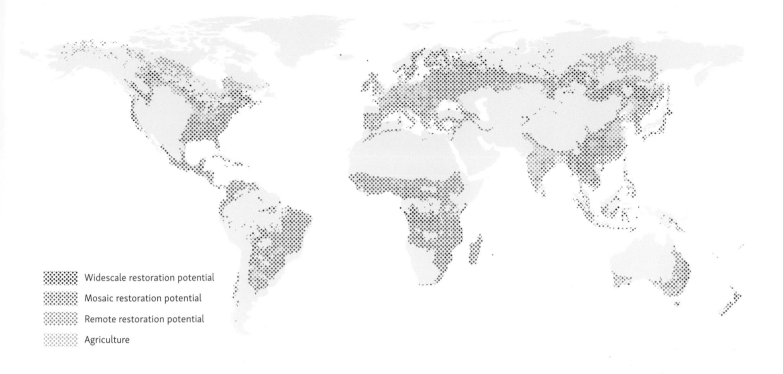

- Widescale restoration potential
- Mosaic restoration potential
- Remote restoration potential
- Agriculture

or bamboo products are considered carbon-neutral. However, if such a product is burned in an electrical power plant (a realistic End-of-Life scenario in Western European countries), the system generates electricity or heat, which can replace electricity otherwise generated from fossil fuels, endowing the product with a carbon credit towards a negative carbon footprint.

Preliminary results (excluding carbon sequestration effect). Adding the cradle-to-gate figures to the End-of-Life credit results in the following main emission components in the carbon footprint of engineered bamboo products (range depending on the product assessed):

- energy consumption for processing: 52–63%
- international sea transport: 15–25%
- local transport (truck): 10%
- use of resin: 3% (flattened bamboo) to 16% (outdoor SWB)

Somewhat unexpectedly, it is not the use of glue or the relatively long transport distance to Western markets but the energy consumption in China (an energy mix dominated by coal energy plants) that has the highest portion in the carbon footprint of engineered bamboo products.[8]

Carbon sequestration through landscape restoration credit

As noted above, so far the carbon sequestration effect has not yet been taken into account. Looking at the highest possible aggregation level (Tier 1 and Tier 2), as described in the IPCC Guidelines for National Greenhouse Gas Inventories,[9] and taking into account changing carbon stocks in forests worldwide, it becomes clear that reforestation on degraded land will lead to a net carbon gain. This provides a tremendous opportunity for bamboo. The World Resources Institute[10] has identified two billion hectares suitable for so-called mosaic restoration, integrating forests including bamboo with

▲ **The World Resources Institute (WRI) identified two billion hectares of degraded land globally that offer opportunities for restoration.**

other land uses such as agroforestry and agriculture. By its prevailing overlaps with the natural growth area of giant bamboo, the restoration map shows the important role bamboo can play to make degraded land productive again, with potentially high yields as an additional benefit.

In INBAR's Technical Report No. 35 on the Chinese situation from 2015, the carbon sequestration benefit for reforesting degraded land with bamboo was quantified and allocated to the carbon footprint of engineered bamboo products. Extra demand of engineered bamboo from China has an effect on carbon sequestration which is similar to that of European and North American wood: it leads to a better forest management and an increase in forest or plantation areas.[11] Because of the increased market demand for engineered bamboo products, the annual growth of Moso bamboo areas in China in the year from 2004 to 2011 has been over 5 per cent, mainly occurring on barren wasteland or poor farming grounds through the Grain for Green programme of the Chinese government.

If the carbon sequestration credit from land use change in China is being included in the carbon footprint of the four main engineered bamboo products, the results demonstrate that their carbon footprint can be CO_2-negative, even when applied in Europe. Apparently the credits for bio-energy production during the End-of-Life phase as well as the carbon sequestration as a result of land change both together outweigh the emissions during production in China and shipping the bamboo products to Europe.

◂ Typical barren grassland which has been rehabilitated with bamboo over the past years.

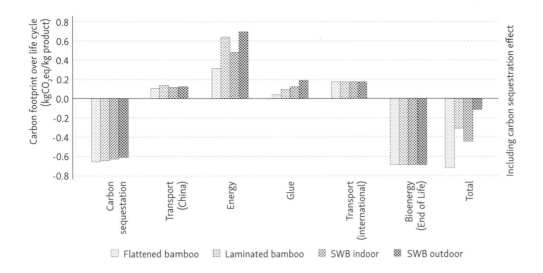

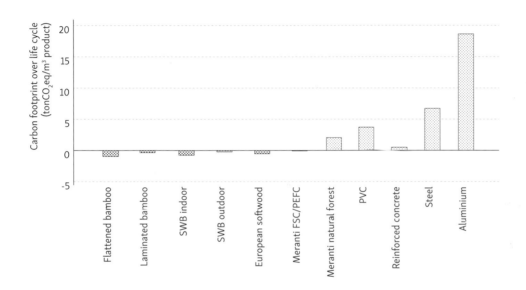

Although the figures do not reflect specific applications – which would include maintenance and material use for required mechanical and functional properties (functional unit) – they give a good indication of how the various materials compare in terms of their environmental impact. This can be used as a basis for more specific calculations for individual applications. Per se, materials pertaining to the biological cycle of the circular economy model have a far lower carbon footprint than energy-intensive materials pertaining to the technological cycle. In terms of carbon footprint, engineered bamboo materials appear to be competitive with sustainably sourced European softwood (which shares a negative carbon footprint as a result of active reforestation and End-of-Life credit for incineration in bio-energy plants), and score better than tropical hardwood, even from sustainably managed plantations (FSC/PEFC). In the case of unsustainably sourced hardwood, including the environmental impact of loss of biodiversity as well as the carbon sequestration debit of clear-cutting, the advantage in favour of engineered bamboo materials becomes overwhelming.

In conclusion, it seems clear that engineered bamboo products with their good mechanical and aesthetical qualities can be a truly sustainable alternative for established building materials, not only from a climate change mitigation perspective (low carbon footprint) but also from the circular economy perspective.[12] If a nutrient from the biosphere can be adequately used as an alternative in the same application as a nutrient from the techno-sphere, it should always be preferred from the sustainability point of view.

▲ Carbon footprint over life cycle (kgCO$_2$eq/kg product).

▼ Carbon footprint over life cycle (kgCO$_2$eq/m³ building material) for various common building materials.

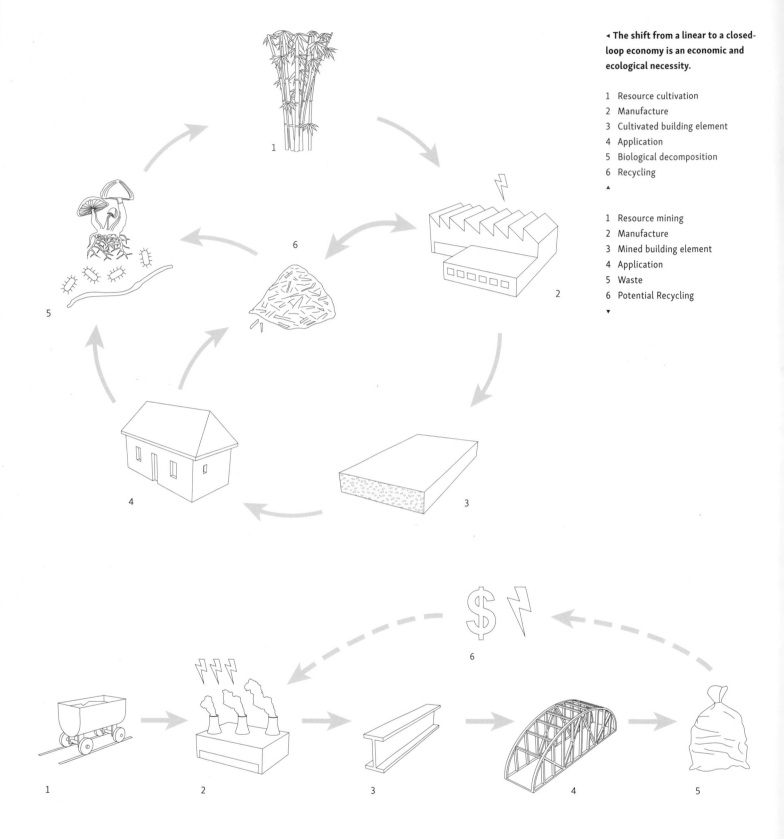

◄ **The shift from a linear to a closed-loop economy is an economic and ecological necessity.**

1 Resource cultivation
2 Manufacture
3 Cultivated building element
4 Application
5 Biological decomposition
6 Recycling
▲

1 Resource mining
2 Manufacture
3 Mined building element
4 Application
5 Waste
6 Potential Recycling
▼

NOTES

1 Webster, K. (2015), *The Circular Economy – A Wealth of Flows*. United Kingdom: Ellen MacArthur Foundation.

2 Ly, P., Pillot, D., Lamballe, P., de Neergaard, P. (2012). "Evaluation of bamboo as an alternative cropping strategy in the northern central upland of Vietnam: Above-ground carbon fixing capacity, accumulation of soil organic carbon, and socio-economic aspects". *Agriculture, Ecosystems and Environment* 149 (2012) 80–90; Rebelo, C. and Buckingham, K. C. (2015), "Bamboo: The Opportunities for Forest and Landscape Restoration", *Unasylva: Forest and Landscape Restoration* 245 (66).

3 INBAR (www.inbar.int).

4 Sharma, B., Bauer, H., Schickhofer, G., Ramage, M.H. (2016), Mechanical characterisation of structural laminated bamboo. Preceedings of the Institution of Civil Engineers – Structures and Buildings, http://dx.doi.org/10.1680/jstbv.16.00061.

5 Rebelo, C. and Buckingham, K. (2015), "Bamboo: The Opportunities for Forest and Landscape Restoration", *Unasylva: Forest and Landscape Restoration* 245 (66).

6 Readers interested in a complete LCA study (ISO 14040/44) – including several other environmental impact indicators – of these engineered bamboo materials are referred to INBAR Technical Report No 35 and the Environmental Product Declarations of MOSO International: Van der Lugt, P. and Vogtländer J. G. (2015), "The Environmental Impact of Industrial Bamboo Products – Life-cycle Assessment and Carbon Sequestration". *INBAR Technical Report No. 35*. Beijing: International Network for Bamboo and Rattan; Agrodome (2016), *Environmental Product Declaration for MOSO solid panel and beam, MOSO Density, MOSO Bamboo X-treme*. Antwerp: CAPEM. Available online via www.moso.eu/epd.

7 The author used LCA software Simapro v8.04, based on Ecoinvent 3.1 as well as the Idemat TU Delft 2015 and 2016 databases.

8 For a more detailed analysis including points for improvements please refer to Van der Lugt, P. and Vogtländer J. G. (2015), see. 6

9 http://www.ipcc-nggip.iges.or.jp/public/2006gl/index.html.

10 Minnemeyer, S., Laestadius, L., Sizer, N., Saint-Laurent, C. and Potapov, P. (2011), *A world of opportunity. The Global Partnership on Forest Landscape Restoration*. World Resources Institute, South Dakota State University and IUCN (available at http://pdf.wri.org/world_of_opportunity_brochure_2011-09.pdf).

11 Lou, Y., Yanxia, L., Buckingham, K., Henley, G., Guomo, Z. (2010). "Bamboo and Climate Change Mitigation". INBAR Technical Report No. 32. Beijing: International Network for Bamboo and Rattan.

12 Cascading from high value-added recycling (OSB or particle board) to lower value-added recycling (burn for green energy production) or just biodegradation.

The following presentations about different kinds of cultivated building materials aim to build the bridge between material research and product application by showcasing selected case studies from various universities, institutes, companies, professionals, and researchers around the world. These materials and products form the spearhead of a new generation of biologically informed resources on the verge of implementation, which will activate their potential for an industrialized and circular building process. The selected examples rely on their respective biological matter in very different processes and to very different degrees, ranging from specific soil and climatic requirements for agricultural cultivation to manufacturing independent of location and soil, as inspired by biological processes.

With a view to this background, the following presentations undertake at the same time a quick journey through several innovative research fields, including but not limited to agricultural sciences, biotechnology, bioengineering, biomimetics, and biomimicry, also showcasing the importance of research in architecture and construction by exploring a new vector for the building industry. It is understood that these case studies represent merely a small selection of the almost limitless possible applications in the field of cultivated building materials. In the quest for the envisioned paradigm they evoke our imagination and creativity shift precisely through their diversity. As such, they represent a springboard for our aspirations and visions.

FROM CULTIVATED MATERIALS TO BUILDING PRODUCTS

SOIL-DEPENDENT PRODUCTS
AGRICULTURE

Today, organic or agriculturally produced materials hold only a niche existence in the material pallet of the building industry. They are even sometimes considered backward-oriented and out-of-date. But this view will change. Architects like Zaha Hadid or Tadao Ando lately realized designs based on traditional organic materials. Wood celebrates a renaissance – even timber soccer stadia are under discussion in the UK and Germany – to a point where the industry is fearing a shortage of resources, so that species such as birch, which have not been used as a construction material so far, are now under research for heavy structural applications in multi-storey buildings. An example is the House of Natural Resources at ETH Zürich at Hönggerberg, where new construction principles of tension-based friction loads in wood construction are tested and applied.

A growing audience of clients is asking questions on material resources, their reuse or recycling, as well as health and environmental risks. This developing awareness opens up possibilities for organically cultivated building materials. At the same time, industries are aware that new products cannot compromise on the level of functionality and develop processes to protect and activate the beneficial qualities of natural resources while adapting their physical behaviour.

A booming market is developing based on agriculturally derived building materials. It is advisable to look critically at products that claim to be 100 per cent biological, as these may have been submitted to chemical treatments which may affect the possibility of full compostation and thus of a closed biological cycle. By contrast, products such as compressed straw boards or mycelium-bond particle boards demonstrate that glues and solvents are not irreplaceable, widening the available material choices in construction.

Agriculture provides two different potential resource streams: through the cultivation of matter like fibres and through a wide variety of agricultural by-products, sometimes seen as waste. As discussed by Peter Edwards and Yvonne Edwards-Widmer in their contribution (see p. 80), the cultivation of building materials through agriculture requires a close examination of advantages and disadvantages in comparison with growing food crop on the land in question. This also applies to considering the use of agricultural waste, which would mean to extract nutrients from a closed-loop metabolism, even if these are carefully used and may be returned to agriculture after serving as building components.

The materials presented in this part tap mostly into so far un- or underused biological resource streams such as algae growth, pulp processing waste, or wasted lignin from agricultural processes – with a surprising variety of architectural applications ranging from insulation foams to structural carbon fibres, from biocomposite façade panels to building bricks. Judging even from this small selection, the hypothesis of growing one's own house seems not as far-fetched anymore.

Wood Foam

Frauke Bunzel and Nina Ritter

Wood foam is the name of a new product developed by the Fraunhofer Institute for Wood Research (WKI) in Braunschweig, Germany. Wood foam is based on lignocellulose, the biomass of the cell walls of trees and other woody plants. Wood foam consists of 100 per cent wood and requires no additives such as binders or adhesives, due to the strength of the material's own bonding forces. Any possible health risks due to emissions from adhesives can therefore be excluded. Furthermore, lignocellulose biomass can be used for production regardless of its quality, which activates a widely available resource. The result of the processing is a lightweight base material with a porous, open-cell structure, low bulk density and highly insulating properties. From an ecological and economic point of view, wood foams provide an interesting starting point for a variety of industrial applications.

Earlier research

A number of attempts to develop foamed insulation materials from wooden components were made in the 1940s. Highly porous fibreboards were produced, based on residues of the sulphite pulp production and alkaline waste liquor using the Ormell-Rosenlund process.[1] In 1944, Wilhelm Klauditz, wood researcher from Brunswick and name giver of the Wilhelm-Klauditz-Institute (WKI),

▴ Wood foams are a lightweight base material with a porous, open-cell structure, low bulk density and highly insulating properties.

▸▸ The production of wood foam involves several process steps to specifically activate the internal bonding capacity, using existing technologies from the wood product and paper industry.

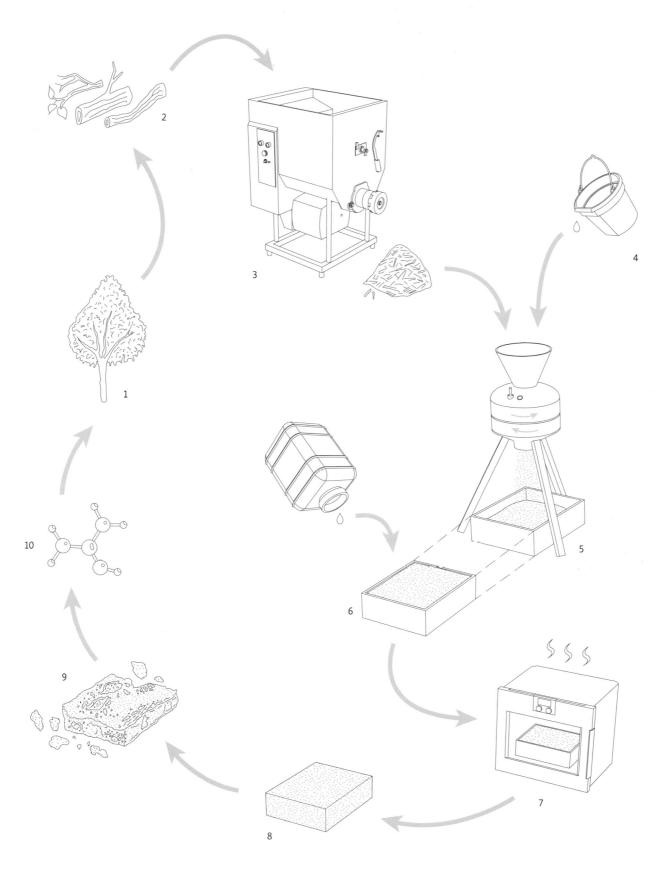

1 Resource cultivation
2 Tree logs
3 Wood chips
4 Water
5 Fibrilation process
6 Foaming agent
7 Thermal treatment
8 Cultivated building element
9 Biological decomposition
10 Biological nutrients

produced wood foam through a hydrated grinding process of fibres. This produced solid wood foam without any additional glue. The strength of the foam depended not on the fibre length but rather on the contact surface of the crossed fibres or fibre fragments. The deforming and hydrating grinding flattened the fibres and made them smooth and slimy, creating optimal conditions for the formation of these surface bonds. The process was used to create light fibreboards from a loose fibre felt of fast-draining long fibres. However, this technology was not competitive with other petrochemical insulating materials, in particular those on a polymer base, and was not further developed. Even the know-how was lost at that time.

Since the beginning of this millennium, the company Innovation Wood (iWood) as well as the University of Natural Resources and Life Sciences (BOKU) in Vienna, and the wood competence centre (Wood K Plus) in Linz have engaged once again in the development of foamed wooden grain materials. The idea for such "wooden bread" originated from a practice used during the Second World War, when sawdust was added to dough in order to save food supplies. iWood has taken up this idea and started baking sawdust, corn flour or bran from grain or rice, yeast, and water to make "wooden bread".[2] In 2004, the University of Natural Resources and Life Sciences took this idea further to develop a porous wood-containing material. Wheat or grain flour is used instead of corn meal, mixed with wooden flour, yeast, natural aggregates, and water to form dough, which is then baked. While the carbohydrates of the wheat flour serve as a nutrient for the yeast, the proteins of the flour give the material

elasticity and consequently an improved gas-holding ability. In a process very similar to the baking of yeast bread, plate-shaped materials are formed.[3] However, this foam has not found an industrial application, due to, among other reasons, the involvement of food products and the related debate over competing land use for food and non-food plants.

Making of wood foam
The development of wood foam at the WKI has followed the approach that the strength of the foam depends less on the length of the fibres but rather on contact surfaces of crossed fibres. A deforming, hydrating grinding process is required to achieve smooth fibres. For high-strength foams also the wood's own bonding strength needs to be activated. This internal bonding capacity originates from the thermal and chemical degradation of hemicelluloses. The various resulting saccharic acids have reactive chemical groups to form bonds between fibres.

The production of wood foam involves several process steps to specifically activate the internal bonding capacity, using existing technologies from the wood product and paper industry. The starting material can be any lignocellulose materials of diverse qualities, among them hard or soft wood, annual plants like hemp and straw, waste from sawmills, and forest thinning, as the raw material is being ground finely. It is important to note that a frequent issue with renewable materials – their qualities or properties differing according to growth regions and environmental exposures – can be neglected here, which constitutes a vital factor

▸ A high degree of fibre fibrillation is considered to be one of the reasons for the mechanical strength of the foam.

▾ Scanning electron microscopic (SEM) picture of a single disintegrated beech fibre.

500 µm

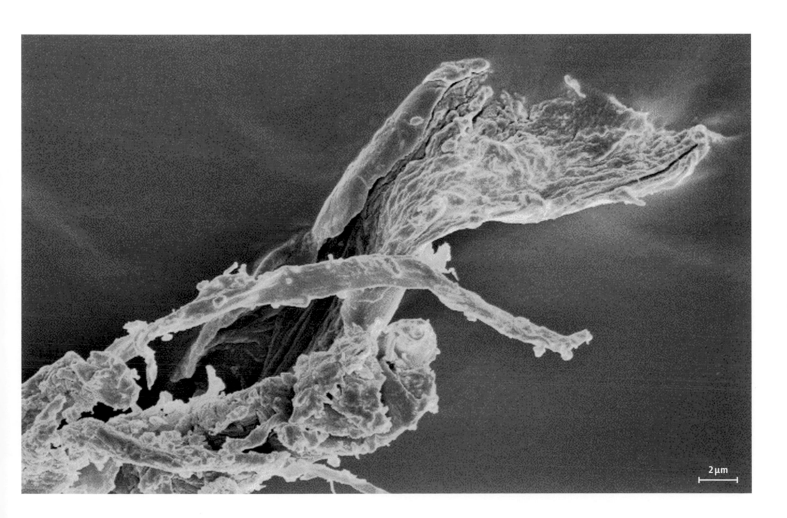

2 µm

▲ Wood foams can be produced in various densities with varying properties.

▼ Wood foams can act as a core material for furniture boards.

for a successful industrialization of a biological material.

The general process is divided into four phases: firstly, the production of wood chips, secondly, their refinement to wood fibres (both are common processes in the wood-based panel industry), thirdly, the preparation of a wood-water suspension with highly fibrillated fibres in a disintegration process, and lastly, the foaming with foaming agents, forming, and drying.

The high degree of fibre fibrillation is considered to be one of the reasons for the mechanical strength of the foam due to mechanical catching, even without any additional additives. The grinding degree of the fibre suspension, more precisely the drainage behaviour, can be determined by the Schopper-Riegler procedure.[4] This method is suitable for a fast control of the grinding process and for the evaluation of the suspensions. In the foaming, three different methods are applicable: using a protein, which is mixed with the suspension and later dried in an oven for several hours; utilizing a

foaming agent, which foams at higher temperatures in the oven; or adding a gas during the drying process. During the drying, the cellulose hardens into a strong structure, wherein the exposed wood components ensure the bonding. No synthetic glues, which could lead to potential health implications due to emissions, are necessary. To shorten the drying time, it is possible to use a high-frequency furnace or a microwave oven.

Properties of wood foam

Wood foam can be produced with different densities from 40 to 200 kilograms per cubic metre. The compressive strength of the foam rises with increasing density and grinding degree and also depends on the original material: the compressive strength of foams made from pine are higher than foams made from beech, due to the longer fibres of pine. The internal bonding strength similarly depends on density and used type of wood.

Thermal conductivity is a key factor for the use of wood foams for insulation. This value depends entirely on the density of the wood foam. Wood

▲ **Wood foams can also be used as packaging material.**

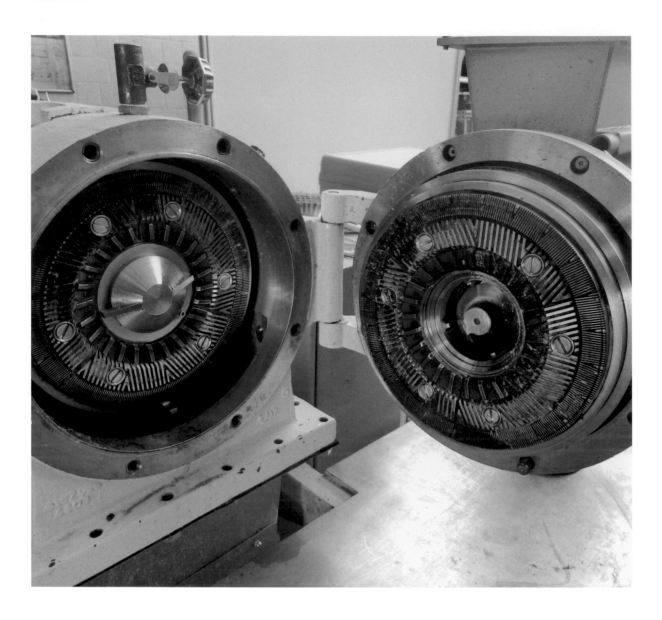

foams with low densities have a thermal conductivity comparable to expanded polystyrene. Because of the open-pore structure, the water absorption of wood foams is very high: essentially, the material behaves like a sponge. However, the structure does not change in terms of swelling or degradation. Also, wood foam from pine fibres shows a lower water uptake due to its high lignin content. The water vapour diffusion resistance factor of wood foam is comparable with mineral wool and the fire behaviour is similar to that of wood fibre insulating materials.[5]

As a result, wood foams have many positive properties including their high strength, low thermal conductivity, dimensionally stable behaviour in water, and a natural flame resistance. The starting material, any lignocellulose material, is a resource available in large amounts and not requiring additional cultivation. Because no chemical additives are necessary, wood foam is recyclable or can be burned. Contrary to certain other wood-based products, wood foam is odourless.

Possible fields of application

One promising potential application of wood foam is the insulation of buildings. The thermal conductivity of foams is low, while their strength is very high. Due to a high water uptake, it is recommended to use wood foams on the façade only as core

▴ The refiner utilizes a deforming, hydrating grinding process to achieve smooth fibres.

materials in sandwich modules. Because of their high compressive strength, it is also possible to use wood foams for footstep suppression, although acoustic tests have not been performed by the time of writing.

Other uses of great potential include the use as packaging material. Sea freight shipment goes along with huge amounts of packaging material, usually in form of plastic foams, generating serious toxic waste problems at the respective destina-tions, as these petrochemical products are usually not recyclable. In contrast, packaging materials made out of wood foam can be treated as biomass for compost, creating nutrients for a circular pro-duction process.

It is conceivable that wood foams can be used as core materials in many applications, for example in furniture production. Different layers can be bonded to such cores, such as wood-based panels, veneers, décor papers, and plastic panels.

NOTES

1 Kollmann, F. (1955), *Technologie des Holzes und der Holzwerkstoffe*. Berlin and Heidelberg: Springer, 579.

2 Affentranger, C. (2003), "Verfahren zur Herstellung fester Stoffe aus pflanzlichem Material, nach diesem Verfahren hergestellter Werkstoff, Verwendung des Stoffes sowie Anlage zur Durchführung des Verfahrens", published patent appli-cation EP 1349949 A1, 8 October 2003.

3 Müller, U. (2013), "Verfahren zur Herstellung von Leichtbauprodukten auf Basis von Holz", published patent application EP 2615209 A1, 17 July 2013.

4 ISO 5267-1 (2000), Prüfung des Entwässerungsverhaltens – Teil 1: Schopper-Riegler-Verfahren (Exam-ination of drainage behaviour – Part 1: Schopper-Riegler procedure)

5 The foams burn and glow, yet the flame goes out partly by itself. Therefore wood foams can be classi-fied as B2 according to DIN 4102.

Biopolymers: Cement Replacement

Leon van Paassen and Yask Kulshreshtha

▲ **CoRncrete samples based on various aggregates.**

Cement production has a significant impact on the environment. With an annual production exceeding 3 billion tonnes,[1] cement consumption reaches an average 500 kilograms per person per year or more than 1,000 kilograms per person per year in urban areas, which is more than the average person eats. The production of cement requires significant amounts of resources (limestone, clay, sand) and energy (mostly coal or lignite) and results in significant amounts of carbon dioxide emissions (about 1 ton CO_2 per ton of cement).[2]

Biopolymers are explored in search of sustainable alternatives to replace cement as a binder in construction materials. Produced by living organisms, biopolymers are composed of small molecules (monomers), which are bonded together by covalent bonds to form larger molecules. Biopolymers are classified in three groups based on their monomer molecules:[3] polynucleotides, like human DNA, consist of 13 or more different small molecules; polypeptides, short polymers of amino acids, are the basic components of proteins; and polysaccha-

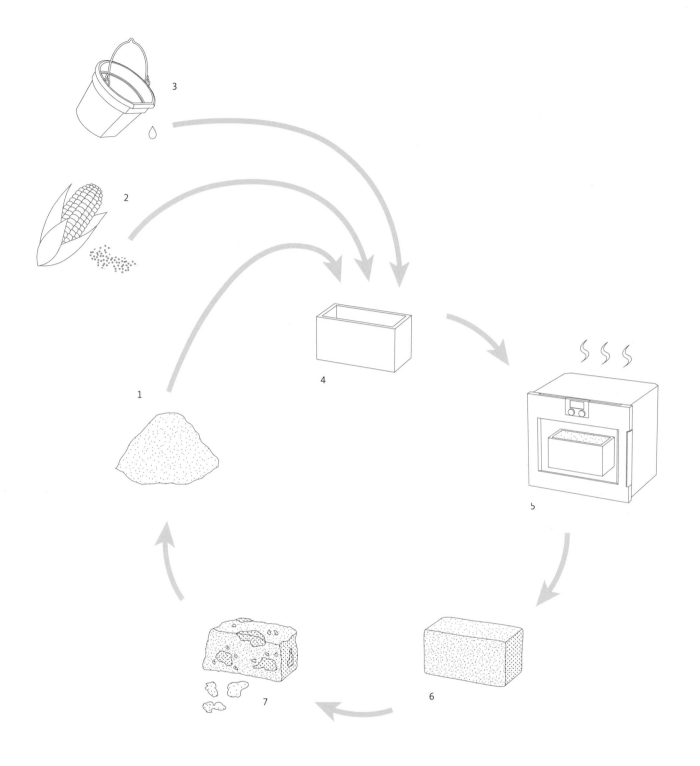

1 Sand
2 Corn flour
3 Water
4 Mould

5 Thermal treatment
6 Cultivated building element
7 Biological decomposition

▴ CoRncrete is a hardened solid material produced by mixing corn starch with sand and water and heating the mixture in a microwave or oven.

rides, like cellulose, starch or alginate, are long linear chains of carbohydrate molecules, such as glucose or fructose.

Biopolymers show a large variety of properties and consequently lend themselves to different applications. Cellulose for example, which is the most common organic compound found on Earth, forms a hard and strong, solid material that provides, as one example, for the structure and strength of wood. Alginate, on the other hand, which is being considered a potential alternative for cement, is water-soluble. The properties of biopolymers depend on their monomers and the amount and type of bonds between these.[4]

In nature, there are many examples where biopolymers act as cementing agents, dating as far back as 4 billion years, to the earliest moments of Earth's existence. In those times, micro-organisms lived in the shallow seas along the old continents. By excreting biopolymers, they attached themselves onto rocky surfaces. This process created thin adhesive films, which trapped suspended sediments in order to form dome-shaped, thin-layered rock structures known as stromatolites. Still today, many organisms use excreted biopolymers to increase their adhesion to surfaces, such as molluscs and oysters in the seas or increase cohesion of soils such as termites and ants on land.

While the main application for natural biopolymers is in food production, their renewability and bio-degradability make them an interesting material for industrial applications as well. The current industrial exploitation in this field focuses mainly on their use in biodegradable plastics as a replacement for polystyrene- or polyethylene-based plastics. Applications of biopolymers in the building and construction industry include their use as a binder in thermally insulating composites, an admixture for viscosity modification in concrete, a modifier in asphalt, or a retarder in the cement hydration process.

▲ At this shore in Hamelin, Western Australia, micro-organisms created thin adhesive films which trapped suspended sediments to form dome-shaped, thin-layered rock structures known as stromatolites.

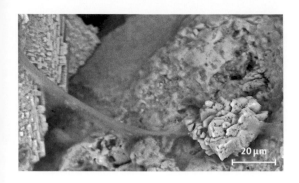

◄ Scanning electron microscopic (SEM) image of biopolymers and calcite crystals grown by feeding nitrate-reducing bacteria within sand.

Biopolymers as cement replacement in construction

The potential performance of biopolymers as a cementing agent can be demonstrated using corn starch.[5] Mixing corn starch with sand and water and heating the mixture in a microwave or oven results in a hardened solid material, named CoRncrete, with unconfined compressive strengths reaching values of up to 20 megapascal. Optimum strength is obtained at a starch-to-sand ratio of 1 to 5 and a water content of 16 per cent.[6] At room temperature, the starch is poorly soluble in water. However, when mixed with sand and water at the optimum ratio, the mixture forms a viscous liquid with self-compacting behaviour. The resulting densely-packed aggregates are one of the reasons for the material's high compressive strength after baking. During heating, the starch molecules dissolve partially and form a gel, which glues the sand grains together and hardens when dried out. Next to the mixing ratio of sand, starch, and water, the strength of the hardened CoRncrete depends on the grain size distribution of the aggregate sand, as well as the heating procedure and time.

Alginate is another biopolymer which can be used as cementing agent. It is found in a wide variety of seaweeds inhabiting temperate as well as cold oceans.[7] The advantage of using alginate is that the seaweed from which it is extracted grows in the sea and does not require land surface to be cultivated. Alginate also differs from starch in that it is soluble in water. However, when mixed with dissolved calcium ions, it will form a gel as the multivalent ions form electrostatic bonds between the long-chained alginate polymers. In trial experiments, dry sand was mixed with 1 per cent alginate powder, percolated with seawater, and simply dried in open air. The resulting sand samples were tested at more than 800 kilopascal.

Biopolymers can also be grown within a material. Percolating a solution of soluble substrates through permeable granular aggregates activates indigenous bacteria to grow and produce inorganic minerals or extracellular polysaccharides (EPS).[8] The application of bio-mineralization in ground engineering applications has been demonstrated at large scale, using microbially induced calcite precipitation (MICP).[9] EPS has not yet been shown to result in sufficient strength improvement to be used as a construction material, particularly in wet environments. However, it has been shown that biofilms can sufficiently strengthen sandy soils to reduce coastal erosion or suppress windblown dust.[10] Besides acting as adhesive component themselves, biopolymers can improve cementation indirectly by controlling the precipitation process of inorganic minerals.

Materials cemented with biopolymers such as alginate or corn starch are strong in compression as long as they are dry. However, once exposed to water, they weaken and disintegrate easily. Further development is required to enhance the durability of biopolymer-cemented materials. Another challenge is the scaling-up to construction applications. Although the temperatures required to heat and dry biopolymer-cemented materials are much lower than for the production of ordinary cement or the fabrication of clay bricks, the heating procedure so far cannot be applied at the scale of a building.

▲ Large-scale biocementation experiment.

▼ Calcit cement can sufficiently strengthen sandy soils to reduce coastal erosion or suppress wind-blown dust.

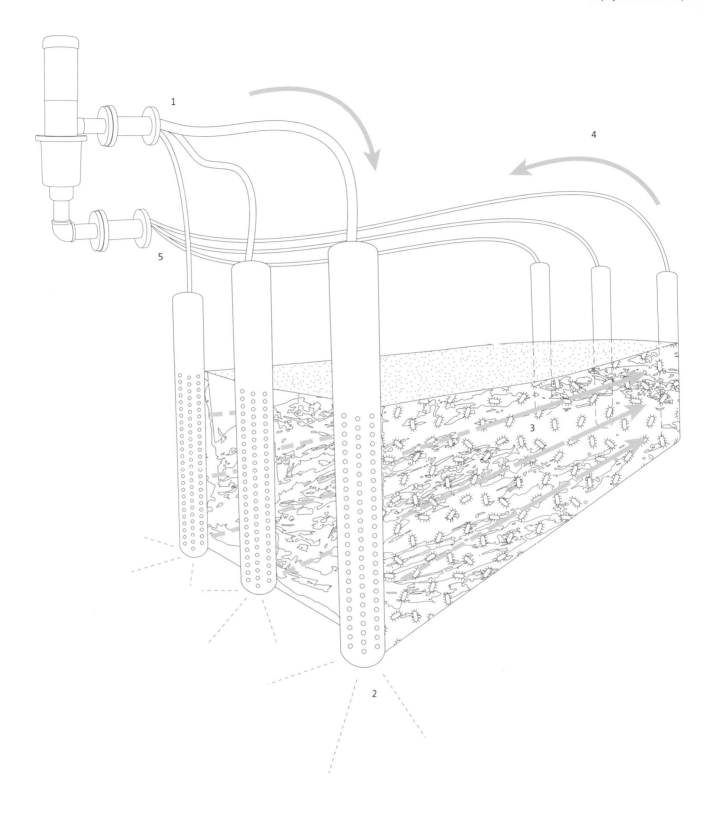

1 Nutrient solution is pumped to the
   injectors
2 Injectors release nutrients into soil
3 Nutrients pass through soil

4 Depleted nutrient solution is
   pumped out
5 Nutrient solution circulates until
   specified values are reached

▲ **Test setup of bio-mineralization
experiment.**

The sustainability of biopolymers generated from agricultural resources such as corn starch is debated, as their cultivation requires a significant amount of land, nutrients, and fertilizers. In that respect, the exploitation of biopolymers from seawater organisms such as alginate seems especially promising. Still, the industrial production of biopolymers for non-food applications competes with the food market and may contribute to rising global food prices. However, recent advances in biotechnology enable the extraction of biopolymers from waste materials, such as the stems and roots of plants, or composted sludge from wastewater treatment plants.[11] Even when combined, the quantities of biopolymers generated from the sea-based cultivation of alginate and an optimal exploitation of land-based organic waste streams will not be sufficient to cover the current consumption of cement. There is no doubt, however, that in combination with other alternative resources and a shift towards a circular building industry, the cultivation of biopolymers is one step towards a sustainable society.

▲ Bio-mineralization in ground engineering applications has been demonstrated at a full scale, using microbially induced calcite precipitation (MICP).

◄ A setup of nozzles and pipes percolates soluble substrates through the ground to activate indigenous bacteria in the ground.

► Full-scale field test installation.

NOTES

1 United States Geological Survey (2011), "USGS Mineral Program Cement Report" (Jan 2011).

2 EIA – Emissions of Greenhouse Gases in the U.S. 2006-Carbon Dioxide Emissions. US Department of Energy.

3 Mohanty, A. K., Misra, M. and Drza,l L.T. (2005), *Natural Fibers, Biopolymers, and Biocomposites*. Boca Raton, FL: CRC Press.

4 Mohanty, A. K., Misra, M., and Drzal, L. T. (2002), "Sustainable Bio-Composites from Renewable Resources: Opportunities and Challenges in the Green Materials World", *Journal of Polymers and the Environment* 10 (1): 19–26.

5 Kulshreshta, Y. (2015), "How to make CoRncrete" – Instruction video https://www.youtube.com/watch?v=hqff5Iq7iCs; Kulshreshtha, Y., Schlangen, E., Jonkers, H. M., Vardon, P. J., and van Paassen L. A. (2017), "CoRncrete: A corn starch based building material", *Construction and Building Materials* (under review).

6 Kulshreshta, Y. (2015), "CoRncrete: A bio-based construction material", MSc thesis Civil Engineering and GeoSciences, TU Delft, The Netherlands.

7 Cybercolloids Ltd., Introduction to alginate, Technical article, http://www.cybercolloids.net/downloads, last accessed on 10 November 2016; Cybercolloids Ltd., The history of alginate chemistry, Technical article, http://www.cybercolloids.net/downloads, last accessed on 10 November 2016.

8 Van Paassen, L. A., Daza, C. M., Staal, M., Sorokin, D. Y., Van der Zon, W., and Van Loosdrecht, M. C. M. (2010), "Potential soil reinforcement by microbial denitrification", *Ecological Engineering* 36 (2), 168–1/5.

9 Van Paassen, L. A., Harkes, M.P., Van Zwieten, G. A., Van der Zon, W. H., Van der Star, W. R L., and Van Loosdrecht, M. C. M. (2009), "Scale up of BioGrout: a biological ground reinforcement method", *Proceedings of the 17th international conference on soil mechanics and geotechnical engineering*, 5–9 October 2009, Alexandria, Egypt; Van Paassen, L. A. (2009), "BioGrout: ground improvement by microbial induced carbonate precipitation", PhD thesis, Delft University of Technology, The Netherlands.

10 DeJong, J. T., Soga, K. S., Kavazanjian, E., Burns, S., Van Paassen, L. A., Fragaszy, R., Al Qabany, A., Aydilek, A., Bang, S. S., Burbank, M., Caslake, L., Chen, C.Y., Cheng, X., Chu, J., Ciurli, S., Fauriel, S., Esnault-Filet, A., Hamdan, N., Hata, T., Inagaki, Y., Jefferis, S., Kuo, M., Larrahondo, J., Manning, D., Martinez, B., Mortensen, B., Nelson, D., Palomino, A., Renforth, P., Santamarina, J. C., Seagren, E. A., Tanyu, B., Tsesarsky, M., and Weaver, T. (2013), "Biogeochemical processes and geotechnical applications: progress, opportunities", *Geotechnique* 63 (4): 287–301.

11 Jiang, Y., Marang, L., Tamis, J., Van Loosdrecht, M. C. M., Dijkman, H., and Kleerebezem, R. (2012), "Waste to resource: Converting paper mill wastewater to bioplastic", *Water Research*, 46 (17), 5517–5530; Van Der Star, W. R. L., Taher, E., Harkes, M. P., Blauw, M., Van Loosdrecht, M. C. M., Van Paassen, L. A. (2010), "Use of waste streams and microbes for in situ transformation of sand into sandstone", Geotechnical Society of Singapore – International Symposium on Ground Improvement Technologies and Case Histories, ISGI'09, 177–182.

Biopolymers and Biocomposites Based on Agricultural Residues

Hanaa Dahy and Jan Knippers

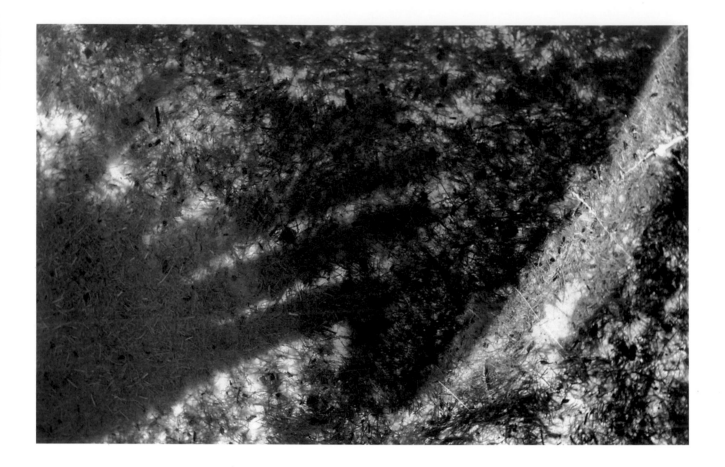

The search for an available, cheap source of renewable, natural fibres as main component in biocomposites and as resource for a number of biopolymers has become an international necessity. Since the late 20th century, global wood consumption has reached a critical level, and the use of crops planted specifically as a source of fibres remains questionable from both social and environmental perspectives. However, the same cannot be said for agricultural residues. Plant lignocellulosic fibres retrieved from agricultural residues such as agro-fibres seem to be a promising alternative resource for efficient biopolymers and biocomposites.

Agro-fibres are available in huge amounts worldwide, mainly as a by-product of the cereal crops harvesting processes. The main interest in annual crop agriculture is the seed itself, the cereal grain,

and not its residues left after harvesting. Consequently, huge quantities of agro-fibres accumulate in the fields. While the global majority is being burned to clear space for replanting, a relatively small amount is used in conventional applications such as bedding, energy applications, or straw-bale housing. The huge volume and low density of straw leads to relatively high transportation costs in comparison to wood. Appropriate densification and safe storage of straw fibres could, however, guarantee the fibres' availability throughout the year and decrease transportation costs. Other types of agro-fibres such as maize or sugar cane do not have the same volume problem, yet are unfortunately not prepared for industrial applications after harvesting either – a fact that highlights a still existing waste management problem of the resource.

▲ TRAshell panels offer different textures and properties depending on the utilized agricultural fibres.

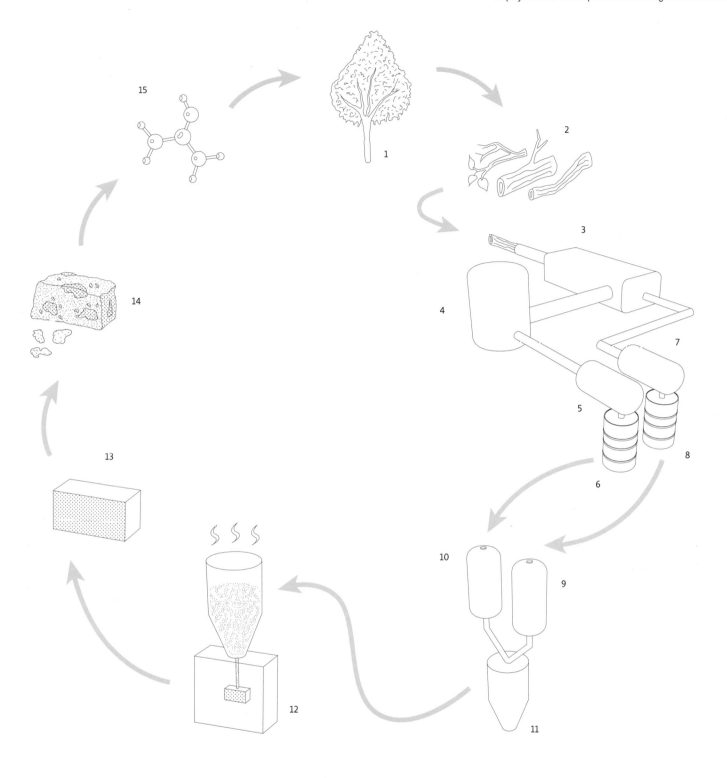

1 Resource cultivation
2 Tree logs
3 Pre-treatment pulping
  for e.g. paper industry
4 Saccharification
5 Fermentation

6 Biofuel
7 Chemical processing
8 Chemicals and monomers
9 Chemical processing
10 Fermentation
11 Biopolymer

12 Thermoplastic processing
  techniques
13 Biopolymer/biocomposite
  building element
14 Biological decomposition
15 Biological nutrients

▲ **Plant lignocellulosic fibres retrieved from agricultural residues are a promising alternative resource for efficient bioplastics and biocomposites.**

Together, those agricultural residues compose a very important lignocellulosic resource, which could replace many wood-based products and serve as a main basis for various chemical and fibre-based industries, depending on their extraction and modification of the celluloses, hemicelluloses, starch, lignin, waxes, fats, minerals, and other materials present in these natural resources. These components can be transformed into biopolymers and biocomposites with potential applications in different industries, including and specifically in the building industry.

Definitions

Biocomposite hereby describes the composition of a fibre and a matrix (binder) where at least one of the two is based on lignocellulosic biomass (i.e. agro-fibres), while *green biocomposite* describes a composite where both main components are lignocellulosic biomass-based. Agro-fibres and thermoplastic polymer combinations in the form of biocomposites are sometimes also referred to as *agro-plastics*.

Biopolymers include both *bioplastics* and *bioresins*. According to the European Bioplastics e.V. association, based in Berlin, Germany, *bioplastic* defines thermoplastic biomass-based polymers, whether biodegradable or not. *Bioresins* are thermoset polymers derived from biomass resources

such as vegetable oils, providing an alternative to existing fossil-based resins that has a lower environmental impact.

Biocomposite and bioplastic applications in architecture

Current applications of agricultural residue-based biocomposites have proven the advantages of natural fibres (agro-fibres) as a basis for a new series of high-quality, recyclable, and environmentally-friendly building materials in contemporary architecture.

Arboskin

In 2013, supported by the European Regional Development Fund (EFRE), a pavilion named Arboskin was constructed on the campus of the University of Stuttgart featuring a bioplastic façade – a hybrid mixture of non-reinforced bioplastics such as polylactide (PLA), lignin, starch, polyester, waxes, and others[1, 5]. The semi-finished sheets served as cladding for flat or freeformed interior and exterior walls. The material can be recycled and meets the high durability and inflammability standards for building materials. The 3.5-millimetre-thick, triangular, thermoformed, and CNC-drilled panels were applied as a façade shell structure, illustrating the possibility to apply pure bioplastics in façade systems in the future.

▲ The Arboskin pavilion on the campus of the University of Stuttgart features a bioplastic façade – a hybrid mix of non-reinforced bioplastics such as PLA, lignin, starch, polyester, and waxes.

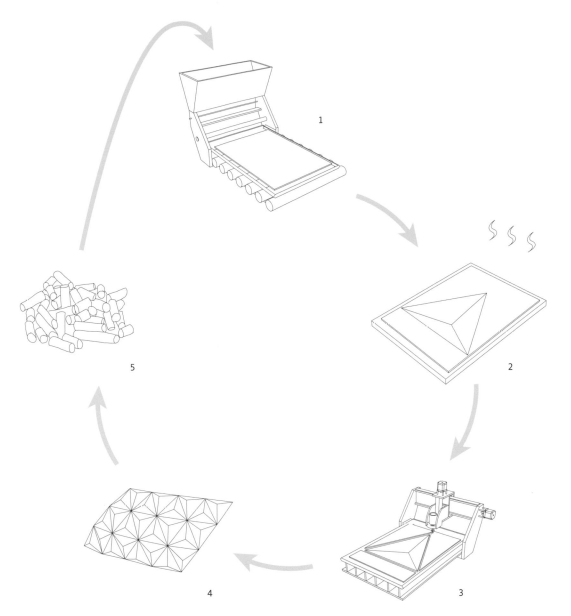

1 Sheet extrusion
2 Thermoforming
3 CNC milling
4 Arboskin application
5 Regranulation

▲ Contour milling allows for multiple variations among the moulding components, as identical thermoformed parts can be processed differently by using various CNC milling paths. The plastic waste that results from the CNC milling process is re-granulated and directly returned to the extruding process.

◄ The freeform bioplastic façade functions as a shell structure based on sheet material, with additional load-bearing and bracing ring carriers and joists.

◄ The structure involves load-bearing properties of the double-curved skin, both in the load-bearing and bracing processes of the entire system.

►► Bio-flexi panels can be twisted and bent accordingly to custom needs, then fixed in their final geometry through veneer layers on both sides.

For the project, collaborating materials scientists, architects, product designers, manufacturing technicians, and environmental experts were able to develop a new material for façade cladding, which is thermo-formable and made primarily (more than 90 per cent) from renewable resources. Developed by project partner TECNARO within the framework of the research project, ARBOBLEND, a special type of bioplastic granules, can be extruded into sheets which are further processable as needed. They can be drilled, printed, laminated, laser-cut, CNC-milled, or thermoformed to achieve different surface qualities and structures. Various moulded components can be produced. The freeform bioplastic façade functions as a sheet material-based shell structure with additional load-bearing and bracing ring carriers and joists. Contrary to common non-load-bearing façade constructions, this structure involves the load-bearing properties of the double-curved skin in the load-bearing and bracing processes of the entire system.

The goal of this project was to develop a maximally sustainable and durable building material while keeping petroleum-based components and additives to a minimum. It demonstrates the potential of modified bioplastics as a bracing material suitable for exterior applications, as this material adds only a minor load due to its own weight. The façade utilizes a minimized number of points of support or mounting brackets on the structural work behind it. The necessary process of contour milling allows for multiple variations among the moulding components, as identical thermoformed parts can be processed differently using various CNC milling paths. This allows for the cladding of

freeform areas with a single moulded component. The plastic waste that results from the CNC milling process is re-granulated and directly returned to the extruding process. At the end of their useful life, the façade sheets can be composted or disposed of almost carbon-neutrally. The ecological audit was completed by the Institute for Water Engineering, Water Quality and Waste Management (ISWA) of the University of Stuttgart. Furthermore, the material's resistance to microbial degradation was determined.

Bio-flexi high-density fibreboards

The developed material[2] is a flexible, recyclable, and compostable high-density fibreboard (HDF) for architectural applications, manufactured from agricultural residue fibres. The raw agro-fibres, which contribute up to 90 per cent to the final material by weight, are bonded without any pre-chemical modification by an environmentally-friendly, thermoplastic, elastic binder using classic plastic-industry machinery.[3] Subsequent compounding through single- and double-screw extrusion manufacturing processes then produces semi-finished Bio-flexi panels. The high-density fibreboards are suitable for the production of freeform furniture and partitions as well as slip-resistant and shock-absorbent floor covering and underlayment.

The fibreboards can be made from wheat, maize, rice, oat, barley, or rye straw. In the case of rice straw, the fibreboards offer an extra advantage due to the high silicate concentration of up to 20 per cent, which allows the material classification "particularly fire-resistant" according to the German DIN 4102-B1 standard by only adding minimal min-

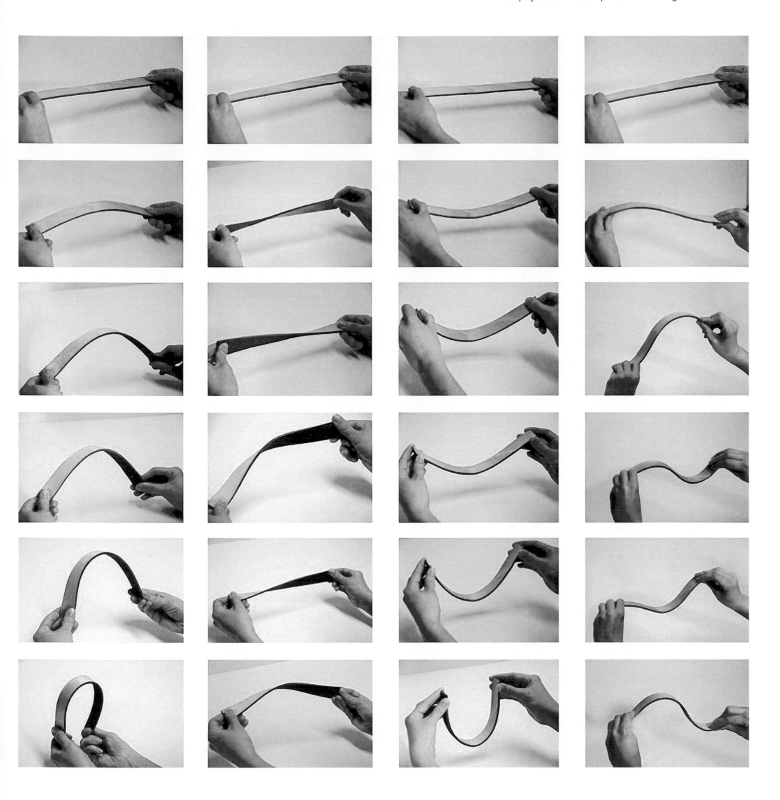

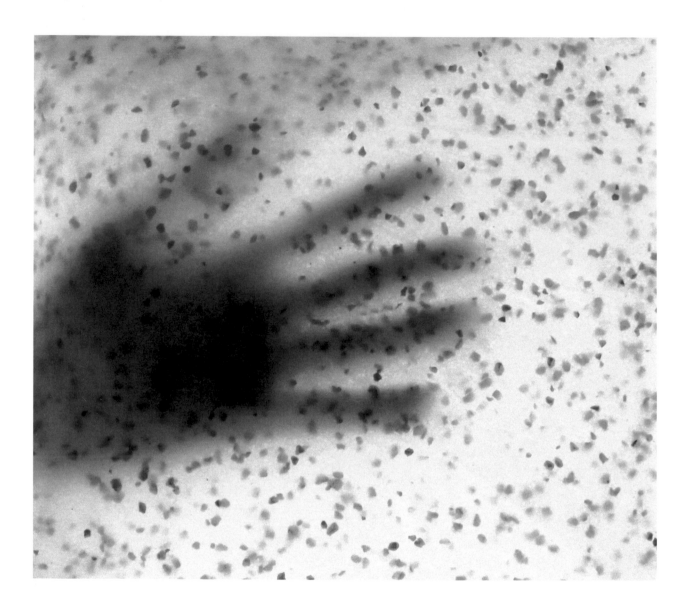

eral-based, flame-retardant additives.[4] These combined parameters minimize health risks during the manufacturing process as well as during the usage life-time when the developed board is applied indoors. The fibreboard has two End-of-Life options: it can be recycled to a number of recycling cycles and consequently industrially composted, which is a highly positive environmental solution that helps minimizing waste accumulation. Such ecologic End-of-Life options are rarely available in the contemporary fibreboards market worldwide.

TRAshell biocomposite façade panels

This prototype – manufactured together with students within an architectural workshop at the Faculty of Architecture and Urban Planning at the University of Stuttgart in 2011 – is a freeform shell of 300 × 300 × 2–5 millimetres, manufactured

through a closed moulding process with green agro-fibre thermoset composites, free-oriented short natural fibres, and a plant-based thermoset bioresin.[5] Different agro-fibres and agro-based particles such as coconut shells, cereal straw, and black coal ash were tested as agro-active fillers within this product design to reach textured effects.[6] The matrix applied is plant oil-based to reach fully green-based biocomposites for architectural cladding applications, thus achieving higher material sustainability in contemporary architecture.

The proposed prototypes were examined for their suitability in exterior façade applications within a doctoral research project at the University of Stuttgart[7] between 2012 and 2014, and the initial free-weathering tests – the samples were subject-

▲ TRAshell panels are manufactured through a closed moulding process from free-oriented short natural fibres, and a plant-based thermoset bioresin.

ed to free weathering on the roof of the building of the Faculty of Architecture and Urban Planning in Stuttgart – indicated their potential success in exterior as well as interior applications.

Printing biocomposites

New technological developments provide an exciting outlook into the future of cultivated building materials. In the short term, extrusion or casting processes allow an additional reduction of weight and enable the integration of additional properties such as thermal insulation into bio-products. Next to these conventional extrusion processes, however, additive manufacturing will change the production, forming, and application of biocomposites and biopolymers in the near future. While 3D printing of PLA filaments is already widely in use, printing in combination with (short) natural fibres as fillers and reinforcement is still at basic research levels. This is sure to be one of the most interesting fields of applying agro-based biomaterials in architecture in the future.

NOTES

1 Köhler-Hammer, C. (2013), "Biobased plastics for exterior facades", *Bioplastics Magazine* 04/13, 12–14.

2 Dahy, H. and Knippers, J. (2016), "Flexible High-density Fibreboard and Method of Manufacturing the same // International Patent pending WO2016005026A1", WIPO/PCT, 14 January.

3 Ibid.

4 Dahy, H. (2014), "Natural fibres as flame-retardants?", *Bioplastics Magazine* 02/14, 18–20.

5 Dahy, H. (2015), "Agro-fibres Biocomposites Applications and Design Potentials in Contemporary Architecture: Case Study: Rice Straw Biocomposites", Forschungsbericht (research report) No. 37. Stuttgart: Insitut für Tragkonstruktionen und Konstruktives Entwerfen, Universität Stuttgart, 2015.

5 Köhler-Hammer, C. (2015), "Anwendungsmöglichkeiten biobasierter Kunststoffe im Innen- und Außenraum von Gebäuden – Beispielhafte Entwicklung 2015", Forschungsbericht (research report) No. 38. Stuttgart: Institut für Tragkonstruktionen und Konstruktives Entwerfen, Universität Stuttgart.

6 Dahy, H. and Knippers, J. (2013), "Product Design Aspects of Agro-Fibres Biocomposites for Architectural Applications", Conference Proceedings of SB13 (Sustainable Building Conference), TU Graz, Austria, 25–28 September, 709–719.

7 Dahy, H. (2015), "Agro-fibres Biocomposites Applications and Design Potentials in Contemporary Architecture – Case Study: Rice Straw Biocomposites", Forschungsbericht (research report) No. 37. Stuttgart: Institut für Tragkonstruktionen und Konstruktives Entwerfen, Universität Stuttgart.

Lignin-based Carbon Fibres

Peter Axegård and Per Tomani

Carbon fibres are an extremely strong and light-weight material and feature in an ever-increasing variety of applications. Today, the demand for carbon fibre is mainly limited by its high cost. Carbon fibre is therefore currently used primarily in high-end applications with a demand for high mechanical performance, such as sports equipment, aircraft, satellites, pressure vessels, specialized tools, wind turbine components, and for reinforcing concrete subject to extreme demands. The introduction of a cost-effective, lignin-based carbon fibre could change the situation and allow carbon-fibre applications in many more segments, while activating the biological, renewable, cultivated resource wood.

Conventional carbon fibre production

The current global production of petroleum-based carbon fibre is approximately 50,000 tonnes per year. The demand keeps growing, even with today's relatively high carbon fibre price. By 2020, the use of carbon fibres is expected to more than double.[1] Worldwide, only a small number of manufacturers produce carbon fibres, dominated by nine companies based primarily in the USA, Japan, and Germany at the time of writing. Structural carbon fibres are mainly manufactured from the petroleum-based product polyacrylonitrile (PAN). A smaller quantity is manufactured from petroleum pitch, which is a by-product from certain oil refining processes, mainly in the USA. As a result

▲ Balsa wood laminate demonstrator with carbon fibre epoxy resin composite; the woven carbon fibre is made entirely from softwood lignin.

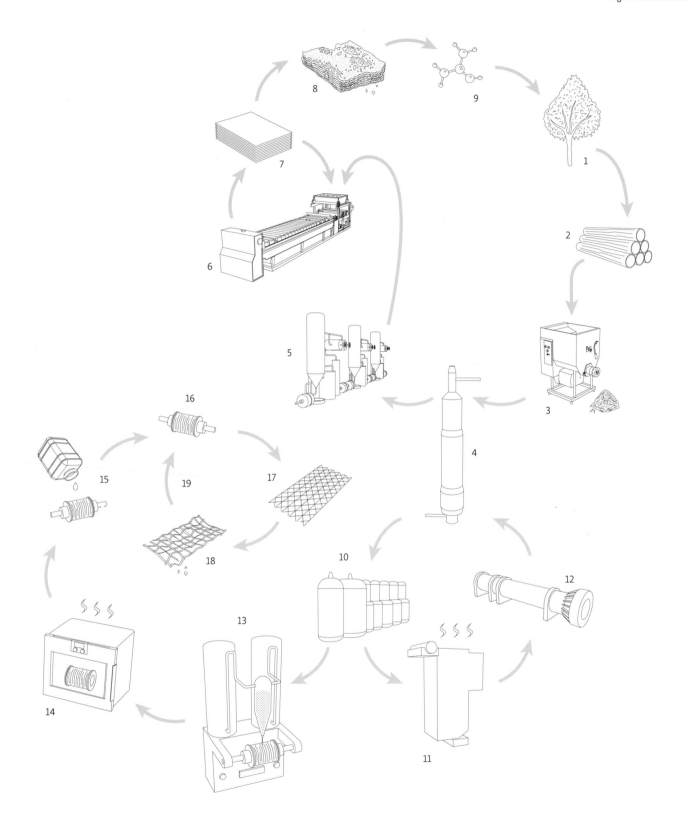

1   Resource cultivation
2   Tree logs
3   Woodchipping
4   Digester
5   Bleaching
6   Paper production
7   Paper product

8   Biological decomposition
9   Biological nutrients
10 Evaporation
11 Recovery boiler
12 Lime kiln
13 Carbon fibre spinning
14 Thermal treatment

15 Surface treatment
16 Fibre spool
17 Carbon fibre application
18 Technical decomposition
19 Carbon fibre recycling loop

▲ The LignoBoost process is able to extract lignin from black liquor, a by-product of kraft pulping. The lignin can be converted into bio-based carbon fibre.

of the large-scale increase of the use of shale gas there, this supply of petroleum pitch is expected to be reduced in the near future.

To produce petroleum-based carbon fibres on an industrial scale, three production steps are necessary. The first stage is the spinning (extrusion) of PAN or petroleum pitch fibres. Thermal processing is then carried out in a number of processing phases: stabilization, followed by carbonization, and, in the case of extremely high-performance carbon fibres, a graphitization procedure. The final stage involves the chemical treatment of the fibre surfaces to make them compatible with polymers. PAN-based carbon fibres are manufactured in a solvent-based process, while pitch-based carbon fibres are manufactured through melt spinning.

The need for an alternative, cultivated approach

In the near future, the automotive industry will be affected by forthcoming specific legal requirements to achieve higher energy efficiency, which will involve a radical technological shift towards alternatively powered cars that can only be reached with new lightweight materials and production technologies. Similarly, the wind turbine of the future, with higher energy performance, will be made possible only through the development of light, stiff, strong, and cost-effective substances. Increasingly, even non-mechanical, non-load-bearing carbon fibre applications are emerging, such as batteries that use commercial carbon fibres as active electrodes, storage of hydrogen gas with activated carbon fibres, or thermal insulation at

◄ Lignin-based carbon fibres link two existing value chains, the forest industry and the carbon fibre industry.

► 100 per cent softwood lignin.

▸ In principle, the same process stages apply for the manufacturing of lignin-based carbon fibre as for polyacrylonitrile-based carbon fibres.

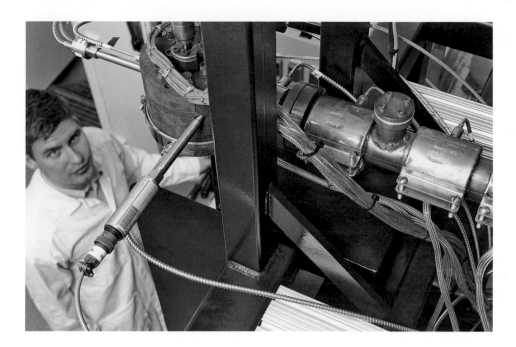

high temperatures. Based on petroleum as a raw material, however, the carbon fibre industry faces rapid growth in demand while relying on a toxic and finite resource. Even using this resource more efficiently would still put the focus onto the wrong starting position. As Michael Braungart once said: "It doesn't make sense to get efficiently to Munich if you want to go to Hamburg."[2]

Swedish research institute Innventia, in cooperation with Chalmers University of Technology in Gothenburg, has developed a process for extracting lignin from the black liquor from the kraft pulp mill process. The innovative process is called Ligno-Boost and has been commercialized since 2008 by Valmet. Two industrial plants have been sold by the time of writing. If properly tailored, the lignin gained by this process can be converted into bio-based carbon fibre. If used solely for this process, the lignin that can be produced in these two facilities alone would amount to a volume that corresponds to more than half of the world's current carbon fibre production. Today, Innventia is able to manufacture continuous multi-filament lignin fibres and use this material to produce carbon fibres that are decimetres in length. Through close examination of each treatment step from black liquor to lignin-based carbon fibres, the parameters of production can be controlled in order to tailor-make different final carbon fibre properties.

In principle, the same process stages apply for the manufacturing of lignin-based carbon fibre as for PAN-based carbon fibres. The production of lignin-based fibre allows a number of simplifications of the process compared with PAN-based fibre and has, therefore, great potential for a more cost-effective and non-toxic process, making it highly competitive. Innventia is currently establishing a pilot facility for the production of continuous lignin-based carbon fibre in order to achieve a larger scale of production and emulate the industrial production of cultivated lignin-based carbon fibre on a large scale.

Composite applications

The project partner, Swedish research institute Swerea, has the expertise and equipment necessary to analyze physical fibre properties linked to various composite applications with standardized test methods for characterizing individual fibres and smaller quantities of fibres. Swerea is able to analyze and verify the process properties of fibres for different manufacturing methods and different matrix systems, and to design, dimension, and develop manufacturing techniques for composites. These skills are essential for the development of alternative fibre materials such as a lignin-based carbon fibre. To move towards a standardized and industrial product, Swerea is building a new testing and demonstration facility, fitted out for the devel-

◄◄ Current research shows that lignin-based carbon fibre can be optimized for both structural (load-bearing) and non-structural applications.

opment and verification of bio-based composite materials.

Together with industry, Innventia and Swerea are planning for a continuous pilot line for lignin-based carbon fibre and ultimately the industrial manufacture of lignin-based carbon fibre composites. Through this process, two existing value chains – forest industry and high-strength fibre industry – are linked together. The driving force for the forest industry is the creation of a new product, and for businesses further down the value chain it is the re-opening of an economic model based on cost-effective, renewable carbon fibres with new players in the field, braking the monopolized situation of the existing market.

Non-structural and structural applications

Current research shows that lignin-based carbon fibre can be optimized for both structural (load-bearing) and non-structural applications. The non-structural applications are probably closer to a

market launch. Potential structural applications for lignin-based carbon fibre include high-performance carbon fibres in wind turbines, aviation, and aerospace, while medium-performance carbon fibres may replace fibreglass in the automotive industry. For example, the requirements in the European Union's End of Life Vehicles Directive (ELV) for a low ash content of residual products could be achieved by replacing fibreglass with lignin-based carbon fibre in composites.

Carbon fibre is an excellent thermal insulator and could be used in the near future for insulation materials in construction and other sectors. For example, if made electrically conductive, carbon fibres have the potential to be used as electrode materials in batteries, in particular in the new carbon fibre battery concept in which the fibre is coated with a thin polymer electrolyte. Lignin-based carbon fibres can also be processed so that they have a very high pore volume, which can be used for storage of hydrogen and other gases.

◄ Carbon fibre is extremely strong, stiff, and lightweight, and it features in an ever-increasing variety of applications.

► The parameters of production can be controlled in each treatment step from black liquor to lignin-based carbon fibres, in order to achieve tailor-made carbon fibre properties.

◂◂ Potential structural applications for lignin-based carbon fibre include the automotive industry but also, potentially, selected parts in wind turbines, aviation, and aerospace.

▸ Spinning of mono-filament lignin fibre from extruded softwood lignin.

▾ Together with industry, Innventia and Swerea are now planning for the manufacture of lignin-based carbon fibre composites in continuous pilot scale (kilogram scale), with the aim to build a first demonstration-scale unit (with an annual capacity of approximately 50 tonnes) or for companies to move directly into commercial scale (approximately 2,000 tonnes/year).

Currently, small amounts of lignin extract from a modern kraft pulp mill can be used as load-bearing carbon fibre components in lightweight car applications. It is clearly more resource-effective to make lightweight materials from lignin as opposed to using it to replace fossil fuel in combustion.[3] On average, less than 1 kilogram of carbon fibre is currently used per car. If this quantity were to increase to 10 kilograms or 100 kilograms per car, this would correspond to the use of 80,000 tonnes and 800,000 tonnes of carbon fibre respectively per year on a global basis. In view of the fact that the global production capacity in 2011 was less than 100,000 tonnes, a shortage could arise if demand increases in a single market area. In this situation, lignin offers an almost abundant cultivated resource for the future growth of the carbon fibre market.

NOTES

1 Roberts, Tony (2012), *The carbon fibre industry worldwide 2011–2020: An evaluation of current markets and future supply and demand*. UK: Materials Technology Publications.

2 TEDx Talks (n.d.), Cradle to Cradle: Michael Braungart at TEDxRhein-Main [video], https://www.youtube.com/watch?v=iY4OeqwKp-c, last accessed on 4 January 2017.

3 Rucks, Greg F., Agenbroad, Joshua N., Doig, Stephen J., and Hutchinson,

Robert A. (2012), "Kickstarting widespread adoption of automotive carbon fibre composites", Rocky Mountain Institute Autocomposites Workshop, Troy Michigan, November 7–9.

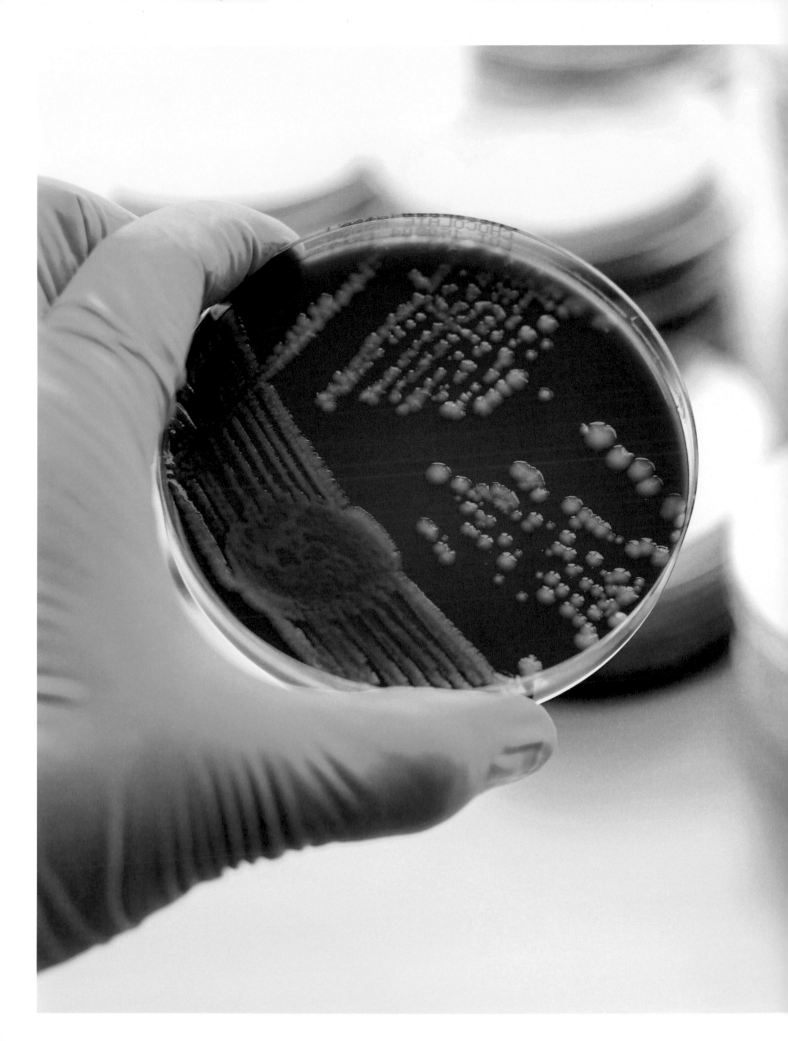

NUTRIENT-DEPENDENT, SOIL-INDEPENDENT PRODUCTS
BIOTECHNOLOGY

Compared to agriculturally harvested building materials, this type of materials encompasses growth and nurturing processes which are soil-independent. The emerging products are not grown in the earth and are therefore not location-dependent. They can be grown where needed, e.g. inside a given structure, as products ready for application which as such do not require further processing.

Biological technology as we understand it here uses natural processes actively to enhance the quality and functionality of products without being harmful for our environment. In this sense, biotechnology-based products have the potential to redefine urban farming and lay the ground for a new understanding of urban economy. They represent a new dimension and new understanding of an urban manufacturing process (as laid out by Dieter Läpple in his contribution, see p. 20).

At the same time, the concept of growing materials inside their applications obviously revolutionizes the architectural planning process. Soil-independent products can potentially be grown inside their site of application, be it a crack in a wall (see contribution by Henk Jonkers, p. 142) or in the city as an entity with its own mode of reproduction (see contribution by Phil Ross, p. 134). While today's building industry is based on the global shipment of a final product, the idea of soil-independent products is to ship the initiator of a growth process, e.g. in the form of bacteria strains or mushroom spores, or at least to achieve steps in between these two extremes. Several entrepreneurs around the world are offering start-up packages of mycelium mixtures that allow the cultivation of home-grown bricks and other building elements or objects. The US-based company Ecovative enables mass-producing chipboard manufacturers to replace the glues in chipboards by mycelium, the cultivated root system of mushrooms. Ecovative took over an existing facility for the fabrication of a product called MycoBoard, demonstrating the low implementation barrier for such an approach: organically grown substances substitute inorganic and sometimes toxic components while leaving the production processes unchanged. Many other potential applications for this technology come to mind in cases where problematic, inorganic adhesives are in use, and on the other hand mycelium is just one conceivable alternative adhesive system, with bacteria, or more precisely the excrements of bacteria, being another.

A new family of 100 per cent organic and compostable materials is developing, where durability and physical performance do not have to be compromised; on the contrary: biological processes can even have a healing capacity, which opens up a new and exciting material classification for architects and engineers. The introduction of new dimensions, such as growth time, into the design and operation of buildings will re-inform the understanding of our discipline.

Fungal Mycelium Bio-materials

Phil Ross

Mycelium is the section of fine, root-like fibres of mushrooms typically hidden from view beneath the earth or within a tree. It can be imagined as an interwoven mesh of thread-like cells that can take on many different forms and consistencies. Fungal mycelium likes to grow on agricultural waste and typically has no other practical value or end use. If there are sufficient nutrients, mycelium will grow quickly and unrestrictedly. The largest known mycelium network is an estimated 2,200 years old and covers a size of about 10 square kilometres in eastern Oregon, USA.[1] The cellular structure of mycelium is defined by a skeleton made of chitin composed of polysaccharides, a feature that distinguishes fungi from vegetal matter (which draws its rigidity from cellulose containing rigid lignin).

Mycelium plays a vital role in our ecosystem, due to its ability to recycle minerals and carbon, as well as in the nitrogen-fixing cycle. Edible fungi are cultivated for a wide range of applications, ranging from human consumption to fungal biological

▲ Mycelium can be used as a binder or adhesive to hold together aggregates and grow any desired object. Dried out, the material is extremely light, yet very difficult to fracture.

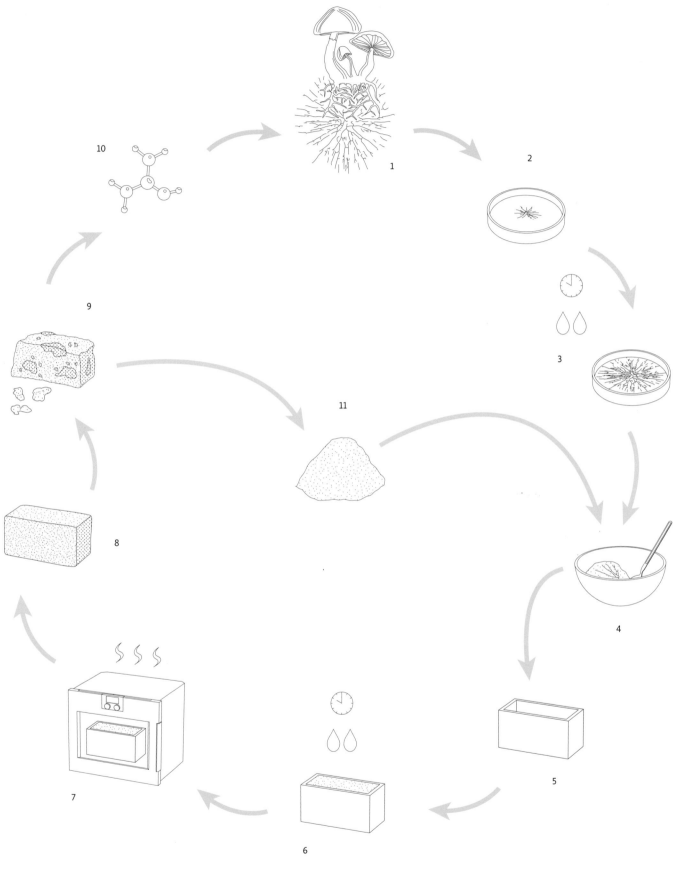

1   Fungal mycelium
2   Hyphae extraction
3   Mycelium cultivation
4   Mixing process
5   Mould
6   Material growth
7   Thermal treatment
8   Cultivated building element
9   Biological decomposition
10 Biological nutrients
11 Substrate

▲ **Growth process diagram of the mycelium.**

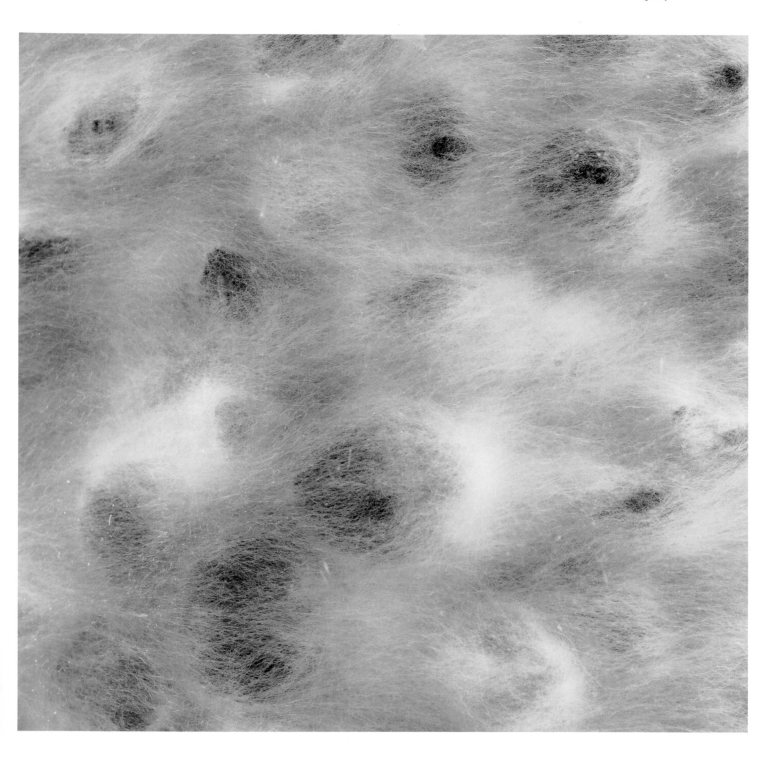

◂◂ The mycelium forms a net-like
structure...
▴ ...which gets denser and denser
during the growth process.

control agents,[2] capture and safe removal of heavy metal contaminants,[3] and even medical uses fighting cancer.[4] In testing, cultivated materials based on fungal mycelium grown on agricultural waste compare favourably to some plastics and other composites made with petroleum-based resins and glues.

Material discovery and intellectual property

My interest and discovery of fungal mycelium in the early 1990s as a material emerged from a life-long interest in biology, technology, and the greater living environment. In those early years, a series of sculptural artworks was created by using living fungus as a primary artistic medium.[5] These artworks were cultivated by infusing live fungal cells into a pulverized, cellulose-based medium (sawdust). The cellulose serves as both food and framework for the organism's growth, and in about a week this aggregate solidifies as a result of the fungi's natural tendency to join together smaller pieces of its tissue into a larger constituent whole. Fungal tissue will bind, solidify, and harden into any chosen form, and once dried out and processed it becomes a lightweight and strong material.

Over the years, these artworks produced out of mycelium materials took shape as a series of geometric forms that could be combined together into suggestive architectural elements. Dried out, the material is extremely light, yet very difficult to fracture. When dropkicked many metres away or smashed against a cement wall, a fungal brick would often bounce back on impact. Under inspection, the bricks did not seem to have any catastrophic failures. They were frangible and cork-like, with a tough, insect-like shell of hardened mycelium skin, and showed an amazing fire resistance as accidental observation demonstrated. This combined empirical knowledge spurred further experiments into the engineering capabilities of materials grown with mycelium and inspired the idea of growing enough bricks to assemble a freestanding, architectural structure. With the help of a mechanical engineer and materials from a local mushroom farm, it was possible to grow several hundred mushroom bricks, and in 2009 these were assembled into a teahouse for an exhibition in Germany.

Around the time this teahouse was growing, several large manufacturers interested in using the material for commercial applications approached the team. It was suggested to apply for a patent to protect the invention and the idea. When looking into the record of existing patents around mycelium materials, it was surprising to see that corporations already owned some of the ideas first shown in artworks. A patent requires a novel proposal that would not be obvious to someone who is proficient in the art being described. The patent would have to have something new in it about making these kinds of materials. The idea of characterizing a defined material with a particular performance

◄ During growth, the mycelium is forming hyphae, which act as tentacles on the search for nutrients.

► Mycelium is an important part of our ecosystem and plays a vital role in the recycling of minerals and carbon and in the nitrogen-fixing cycle.

▸ Different nutrients can be used to feed the mycelium.

▾ A mycelium-based object, grown by MycoWorks Inc. Fungal bricks bounce back on impact.

outcome posed problems in respect to fungal growth. There are many ways to alter the quality of fungus-based materials as they are growing, and many variables that can be used to push/pull the organism towards a desired outcome. Mycelium materials can be grown to be soft or hard, light or heavy, strong and durable, or weak and fracturing. Or, any combination of these characteristics, and others too. Mycelium is like plastic and can be grown to express a number of different qualities within a single object. The patent, in the end, reflected a process for making mycelium materials that are light and strong, and best for use in architecture and design applications for interiors. The patent was granted after a six-year application.

Measuring the unmeasured

Some of the cultivated mycelium materials were strong and light, but it was unknown how strong they actually were or how to describe their qualities other than through vernacular descriptions: tough, springy, brittle, rubbery, airy, etc. Consequently, Hari Dharan, Professor of Mechanical Engineering at the University of California, Berkeley, who is a world expert on composites, started working on the project with one of his PhD candidates, Sonia Travaglini. They began characterizing mycelium composites in a rational and ordered process and started by measuring a control, a form of the material which could be grown repeatedly and consistently, serving as a base of information to expand from. The results showed that the control behaved a lot like expanded polystyrene. Test results and publications demonstrate that mycelium composites are naturally flame-resistant, good insulators, and free of toxic volatile

organic compounds (VOCs).[6] Mycelium-based bio-materials can achieve compression, thermal and acoustic insulation, and flexural characteristics similar to petrochemical composites and engi-neered wood. Research started to allow comparing mycelium with wood, stone, plastic, and other common elements in the constructed world.

Towards application

As the interest in the material qualities of fungi grew, the initial ideas were not motivated by any practical application. The emerging experiments with growing architectural materials out of myceli-um were motivated more by ideas from philosophy than from engineering. In the course of consider-ations about how the material's characteristics might be demonstrated, in collaboration with The Workshop Residence in San Francisco, a series of locally grown mycelium chairs were designed and produced. This was shortly followed up by a com-mission from the San Diego New Children's Muse-um to create an exhibit that included a room full of 1,200 building blocks grown to order. These mycelium blocks were created for use in open-end-ed, hands-on play similar to Legos. When the exhi-bition closed in the fall of 2014, the museum was able to add all of the bricks into San Diego's munic-ipal compost programme, leaving zero waste.

With a patent pending and successful demonstra-tion of the material in the form of the chairs and children's building blocks, it seemed like the appro-priate time to form the technology into a business. MycoWorks was incorporated in 2013 with three co-founders, and a first task was to figure out the kind of business the team wanted to be, as well as finding a way of how to manufacture the materials. While continuing the iterative engineering tests and experiments, the new company started to look at the opportunities and potential for organic com-posites. Research was spent on fibre-based panels production and costs for scaling this production so as to achieve competitive pricing within the market constraints of shortages, tort, shipping, and other pressures altering the global costs of manufactur-ing. MycoWorks plotted out how to alter a giant mushroom farm so it might produce the desired materials. This would require an enormous factory footprint of 30,000 cubic metres and an enormous budget of 60 million US Dollars to get to that scale of production.

MycoWorks continues to experiment with materi-als that will have greater value, yet can be generat-ed with less resources and real estate. Mycelium composites could replace petrochemical foams and other polymer-bound materials used in the build-

▲ MycoWorks grew 1,200 building blocks for the San Diego New Chil-dren's Museum as part of a bigger exhibition. The blocks were compost-ed after the exhibition closed its doors.

ing, automotive, geotextiles, and maritime industries. It is difficult and expensive to bring a new material into the marketplace. For mycelium to be accepted by industry it will have to do one thing very well at a reasonable cost. After that, it will open to its own multitudes.

▲ Mycelium composites could replace petrochemical foams and other polymer-bound materials used in the building, automotive, geotextiles, and maritime industries.

NOTES

1 Stamets, P. (2005), *Mycelium Running: How Mushrooms Can Help Save the World*. Ten Speed Press.

2 Shah, P. A. and J. K. Pell. (2003), "Entomopathogenic Fungi as Biological Control Agents", Applied Microbiology and Biotechnology, 61(5-6):413-423.

3 Galli, U., Schüepp, H., Brunold, C., (1994), "Heavy Metal Binding by Mycorrhizal Fungi," *Physiologia Plantarum*, 92(2):364-368.

4 Wu, J. Y., Zhang, Q. X., Leung, P. H., (2007), "Inhibitory Effects of Ethyl Acetate Extract of Cordyceps Sinensis Mycelium on Various Cancer Cells in Culture and B16 Melanoma in C57BL/6 mice", Phytomedicine, 14(1):43–49.

5 These first undertakings were done by the author of this article, Phil Ross, co-founder and Chief Technology Officer of Bay Area startup Mycoworks Inc.

6 Travaglini, S., Noble, J., Ross, P. G., and Dharan, C. K. H. (2013), "Mycology Matrix Composites". In: Bakis, Charles E. (ed), *Proceedings of the American Society for Composites – Twenty-eighth Technical Conference*. Lancaster, PA: DEStech Publications; Travaglini, S., Dharan, C. K. H., and Ross, P. G. (2014), "Mycology Matrix Sandwich Composites Flexural Characterization". In: Kim. H. (ed.), *Proceedings of the American Society for Composites – Twenty-ninth Technical Conference*. Lancaster, PA: DEStech Publications; Travaglini, S., Dharan, C. K. H., and Ross, P. G. (2015), "Thermal Properties of Mycology Materials". In: Xiao, X., Loos, A., and Liu, D. (eds.), *Proceedings of the American Society for Composites – Thirtieth Technical Conference*. Lancaster, PA: DEStech Publications.

Limestone-producing Bacteria: Self-healing Concrete

Henk M. Jonkers

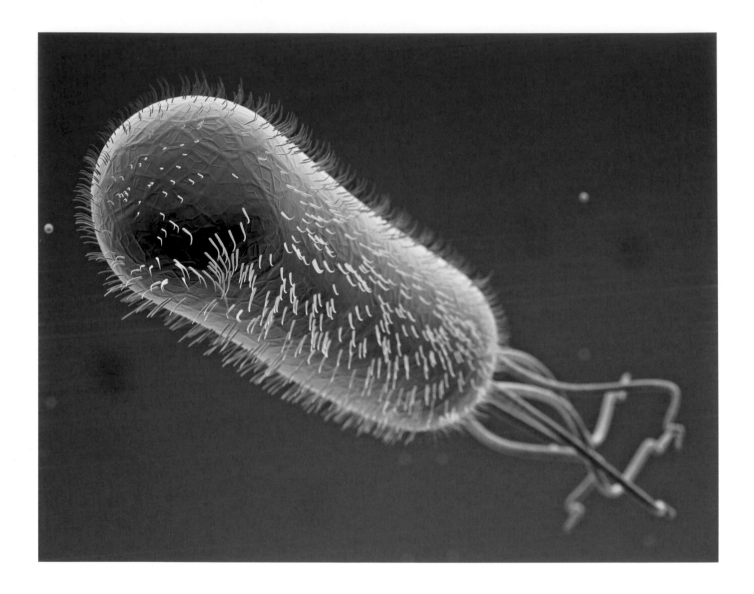

Man-made constructions are typically designed according to the concept of damage prevention. In order to ensure structural integrity and prevent repairs during service lifetime, constructions are often over-dimensioned in terms of quantity of building materials used. Consequently, these robust constructions require more primary materials than strictly needed – what may result not only in high costs but also high environmental pressure.

Nature uses a radically different design approach, one that can be defined as damage management.[1]

In the latter design concept, a certain level or amount of damage is allowed as the living material is able to self-repair occurring injuries. This intrinsic material property allows nature to build elegant structures, in which the minimum amount of material is applied in highly functional constructions.

Living materials featuring such in-built, self-healing properties are endowed with the potential to adapt to often-changing environmental conditions. A tree branch, which increases in weight during growth, can form more structural support tissue exactly at

▲ **Carbonate skeleton produced by the single-celled Foraminifera calcarina sp.**

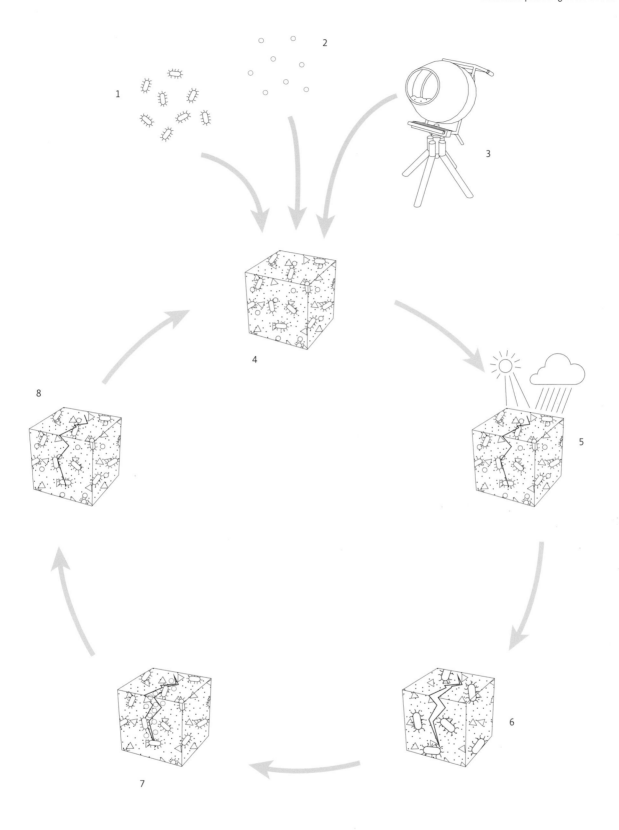

1 Bacteria
2 Nutrients
3 Concrete
4 Self-healing concrete

5 Concrete cracking through
   environmental influences
6 Bacteria activation through
   exposure to environment

7 Bacteria-induced limestone
   precipitation
8 Healed concrete

▲ **The robust, naturally occurring
bacteria – either *Bacillus pseudofirmus*
or *Sporosarcina pasteurii* – already
exist in highly alkaline lakes and can
remain dormant for up to 200 years.**

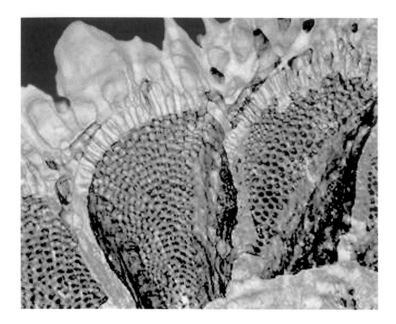

the place where additional strength is required. Tissue damaged in an accident due to temporary overload or sudden impact can restore itself and even improve material properties in such a way that it will be protected against similar stresses in the future.

The challenge of modern designers and engineers is to develop man-made materials which feature such nature-based damage management properties. These novel materials will lay the groundwork for elegant designs which require smaller amounts of materials than classical man-made structures. Such constructions are not only potentially more economical but also more sustainable, as less primary materials and energy in both construction phase and functional service life are required.

A beautiful example of an elegant and material-saving limestone construction design is the skeleton of one-celled Foraminifera. These unicellular protozoa precipitate calcium carbonate inside the cell in a controlled fashion to produce a skeleton that provides structural support and shelter.

Bio-adaptive materials from limestone-producing bacteria

In a number of studies geared at cultivating such a nature-based material at the Delft University of Technology in The Netherlands, bacteria which, under specific conditions, are able to produce limestone, were added to concrete mixes in order to make the material self-healing. Concrete construc-

tions are traditionally made robust in order to prevent cracking, which would result in a lack of water tightness and a consequent early corrosion of the embedded steel reinforcement. By allowing to reduce the thickness of the protective concrete layer, self-healing concrete would enable the design of much more elegant and low-material-based constructions. Occurring cracks would heal themselves, resulting in functional water tightness and effective protection of the embedded steel against corrosion.

Typically a few micrometres in length, bacteria constitute a large domain of prokaryotic micro-organisms. Among the first life forms to appear on Earth, bacteria still today are present in almost all of its habitats. Typically, about 40 million bacterial cells can be found in a gram of soil; up to a million bacterial cells inhabit a millilitre of fresh water. Together, the approximately 5^{1030} bacteria[2] on Earth constitute a biomass exceeding that of all plants and animals.[3] To be precise, only a small percentage of bacteria so far has been even identified and categorized – with many more discoveries waiting. Well known, however, is the importance of bacteria as a vital element of the planet's nutrient cycle and the fact that many stages of the process are dependent on these organisms. Bacteria therefore also live in symbiotic and/or parasitic relationships with plants and animals.

Today, diverse industries are cultivating bacteria in countless industrial applications, such as sewage treatment, the breakdown of oil spills, the produc-

▲ Mode of operation of self-healing concrete.

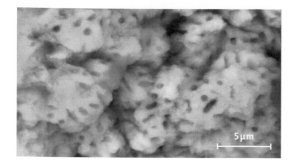

tion of cheese and yogurt through fermentation, the recovery of gold, palladium, copper, and other metals in the mining sector,[4] as well as the huge fields of biotechnology and medicine, for example in the manufacture of antibiotics.[5] So far, this immense resource has not been made productive for the construction industry.

Specific bacteria produce limestone, or calcium carbonate, by metabolically converting calcium- or carbon-containing nutrients. This metabolism changes the local concentrations of calcium (Ca^{2+}) and carbonate (CO_3^{2-}) ions, resulting in the precipitation of limestone, or calcium carbonate ($CaCO_3$). The process of limestone precipitation occurs fast under alkaline conditions, a typical environment in concrete.

One approach to cultivate high amounts of bacterially produced limestone is to provide the chemical compound urea to selected bacteria or derived enzymes. This nitrogen-containing compound can be enzymatically hydrolyzed by bacteria, a process during which carbonate ions are produced. However, as this approach also produced the double amount of environmentally unfriendly ammonia, this metabolic pathway is unrecompensed for sustainability reasons. An alternative to urea are organic carbon compounds such as organic calcium salts, which under alkaline conditions in concrete can be converted to limestone.

Prototypical testing and application of self-healing concrete

Self-healing concrete is made by adding a self-healing agent to the conventional concrete mix. The agent consists of bacterial spores, which are dormant bacteria, and nutrients such as organic calcium salts in encapsulated form. The dormant bacteria become activated by crack-ingress water and subsequently convert the nutrients to limestone. Water entering the cracks functions as the

◄ **Limestone produced by bacterial metabolic conversion of organic calcium salts. Imprints of bacteria in form of dark 2-μm-sized rods are clearly visible within the mineral matrix.**

► **Concrete self-healing agent (dormant bacteria and nutrients) in concrete.**

▼ **Self-healing concrete exhibited as part of Daring Growth at the Biennale, 15th International Architecture Exhibition, in Venice in 2016.**

activator of the system, and as such the concrete actively responds to environmental stresses (crack formation). In new constructions, the healing agent can be added to the concrete mix to make the material self-healing. In existing constructions, the agent can be applied in form of a water-based liquid or in form of a repair mortar or coating to deliver self-healing properties from the outside.

To test the earlier laboratory results in full-scale applications, Delft University of Technology in col-

laboration with Waterboard Limburg (Waterschaps-bedrijf Limburg) and prefab producer Bestcon designed and built a wastewater treatment tank composed of prefab steel-reinforced concrete elements. The tank was placed in spring 2016 and taken into full service in autumn of the same year. Several elements of this tank were made of self-healing concrete, while the remaining panels were made of traditional concrete. The aim of this demonstrator project is to analyze the self-healing capacity of bacteria-based concrete in comparison

▲ A concrete self-healing agent contains dormant bacteria and nutrients in encapsulated form. The picture shows a compressed powered-based agent.

▾ Concrete self-healing agent based on expanded clay particles.

to traditional concrete and quantify the functional and economic performance over the construction's service lifetime.

A specific type of self-healing concrete has been developed and applied in irrigation canals for the town of Ambato in Ecuador. The initial problem with the existing earthen wall concrete liner is that soil movements result in the formation of cracks, threatening subsequent soil erosion and the possible collapse of the irrigation canals. In order to minimize otherwise necessary repairs and to avoid a decreased functionality of these canals during periods of repair, plant fibre-reinforced, self-healing concrete was applied. The incorporated natural fibres derived from the Abacá plant provide a sufficient tensile strength in the concrete liner to avoid the formation of large (millimetre-wide) cracks. Additionally, the integrated bacteria-based self-healing agent provides the necessary capacity to seal and waterproof occurring micro-cracks.

Another application of a bacteria-based healing agent suspended in a water-based liquid was completed on an existing thin, roof-bearing concrete shell designed by architect Frank Marcus. Shrinkage cracks, which had occurred in the concrete cover, were effectively sealed with limestone. Bacteria produced the sealant by metabolic conversion of organic calcium salts in the liquid system.

Economics and sustainability

Bacteria-based self-healing concrete represents a novel class of materials, which potentially allows to design and build elegant constructions which require much less material than traditional concrete constructions. The intrinsic self-healing properties of this hybrid and living concrete make it a material that allows to design according to the damage management concept typical for living natural materials. The application and cultivation of such a material not only allows elegant designs but also saves use of primary resources, making it both an economical and sustainable construction material.

▲ Wastewater treatment tank composed of both self-healing and traditional prefab concrete elements to observe and measure the new material's self-healing capacity over time.

NOTES

1 Van der Zwaag, Sybrand (ed.) (2007), *Self Healing Materials*, Springer Series in Materials Science. Dordrecht: Springer Netherlands.

2 Whitman, W. B., Coleman, D. C., and Wiebe, W. J. (1998), "Prokaryotes: the unseen majority", *Proceedings of the National Academy of Sciences of the United States of America* 95 (12): 6578–83.

3 Hogan, C. Michael (2010), *Bacteria. Encyclopedia of Earth*, eds. Draggan, S. and Cleveland, C. J. Washington, DC: National Council for Science and the Environment.

4 "Metal-Mining Bacteria Are Green Chemists", *Science Daily*, 2 September 2010.

5 Ishige, T., Honda, K., and Shimizu, S. (2005), "Whole organism biocatalysis", *Current Opinion in Chemical Biology* 9 (2), 174–80.

Microbially Induced Calcium Carbonate: Multilayer/Multimaterial Coatings on Non-conductive Materials

Filipe Natalio

Biominerals are minerals formed by organisms. The most prominent examples of such biominerals are bone, nacre, corals, sponge spicules, and shells of diatoms and radiolarian. Biomineral formation – so-called biomineralization – belongs to one of the most complex and yet fascinating processes driven by living organisms and is the end-product of a well-orchestrated coordination between inorganic and organic materials, driven by specialized cells and following a brick-mortar assembly principle.[1] Since their discovery, biominerals have inspired several generations of scientists, engineers, architects, and designers.

Biominerals combine an inorganic component with an organic part to yield composite materials that are "intelligently" distributed in a three-dimensional space.[2] Take bone as an example: bone is an organic-inorganic composite where the organic part is comprised of collagen, water-soluble and water-insoluble proteins, and the inorganic part consists of carbonated hydroxylapatite. However, if the mixture was made under controlled laboratory conditions (also using cells), it would never be classified as bone. Why is that? Only the combination and distribution of this composite material in a three-dimensional space provides bone its unique-

▲ Growing bacterial-induced single-layer $CaCO_3$-coating on polymeric surfaces before and after immersion into the bacterial "cocktail". The white layer is attributed to $CaCO_3$ polymorph vaterite.

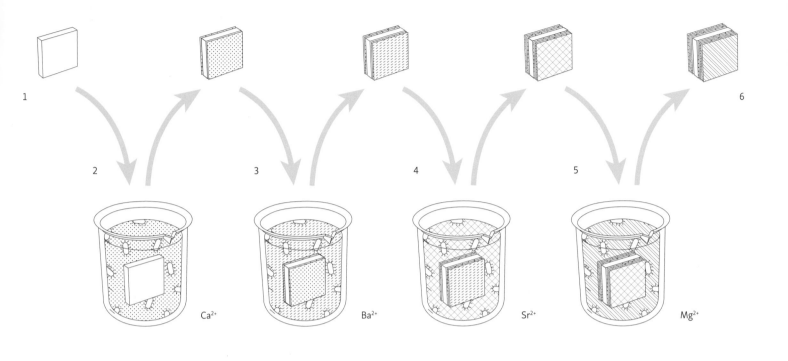

1   6

2   3   4   5

Ca^{2+} Ba^{2+} Sr^{2+} Mg^{2+}

1   Uncoated building element
2   Bacteria-induced CA^{2+} coating
3   Bacteria-induced BA^{2+} coating

4   Bacteria-induced Sr^{2+} coating
5   Bacteria-induced Mg^{2+} coating
6   Coated building element

▲ **Schematic representation of the sequential application of the dip-coating approach for biologically controlled growth of $CaCO_3$-based multilayers on the surface of non-conductive polymers. This approach uses either always the same or various types of different alkaline earth-metals ions (as drawn – calcium, barium, strontium, and magnesium) chemistry to yield the correspondent solid carbonate layer to enable the growth of multi-material coatings.**

ness and superior (mechanical) properties when compared with its pure inorganic or composite counterparts. This concept is valid to all biominerals, independently of their chemical nature (e.g. silica, calcium carbonate, carbonated hydroxylapatite). Biominerals are therefore an excellent example of the combination of materials and structure and have been attracting much research, as they hold a great promise for designing and cultivating future generations of growing materials.[3]

As a general rule in materials sciences, the function equals materials plus structure (function = materi-als + structure). This can be represented as a *function-balance* where material is on the one plate and structure on the other. Where in current approaches, and driven by the industrial revolution mindset, the balance weighs more on material, biominerals (and natural materials) put more emphasis on structure. The reasons are obvious: material availability, (bio)energetic costs, and survival. The natural process uses the available material, produces few building blocks with significant gains in terms of energetic costs, and creatively assembles the same building block in different combinations – the so-called *hierarchical assembly/*

▶ Scanning electron microscopic (SEM) images of biominerals with different morphologies and chemical composition, yet the same underlying principles: $CaCO_3$-based biominerals:
a) nacre, b) coral sclerites,
c) bacterial-induced precipitation.

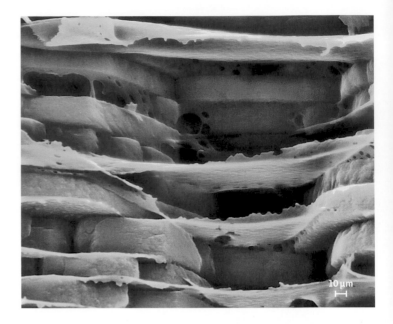

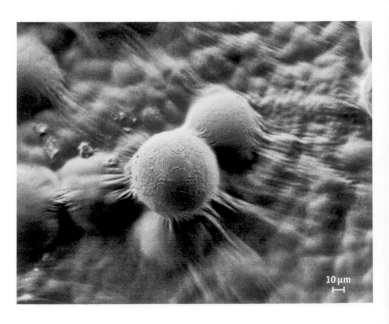

▸ **Silica-based biominerals:**
d) sponge spicules, e) radiolarian,
f) diatom

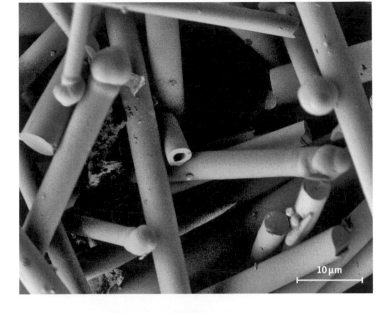

organization – yielding materials with identical or similar compositions but different properties: the ultimate functional material.

Both concepts of *form follows function* and *hierarchical assembly/organization* of building blocks have been explored to translate this embedded information on function towards the new generations of functional biominerals.[4] Despite the great body of work in the field of biomimetic and new bio-inspired solutions, materials and structures today are generally thermodynamically stable.[5] A table of wood and metal, for example, will always stay a table and only be affected through time. Thus, it displays a static function. On the other hand, and without entering non-equilibrium thermodynamic considerations,[6] biological systems are dynamic with a cycling flow of energy. Biological systems that produce materials such as biominerals dispose of a dynamic process of growing mate-

rials. This process is worth exploring when considering the concept of cultivating functional materials.

Particularly interesting are micro-organisms, e.g. bacteria, that are able to produce calcium carbonate ($CaCO_3$) under alkaline conditions – microbially induced calcium carbonate precipitation. The use of $CaCO_3$-precipitating bacterial strains for (bio) cementation processes has been extensively described for a wide range of applications such as self-remediation and crack prevention, as shown in Henk M. Jonkers' contribution to this book (see p. 142), dust/sand/soil fixation in mining sites (described by Leon van Paassen and Yask Kulshreshtha in their contribution on p. 108), or fabrication of masonry and architectural concepts.[7] However, one application of microbially-induced calcium carbonate that remains fairly unexplored is its use for coatings on non-conductive materials (e.g. polymers). Unlocking the possibility to control the biological growth of coatings on non-conductive materials would envision a new technological and/or large-scale application.

Cultivating coatings

Coatings play an important role in modern materials across a multitude of scales – from nano (10^{-9} metres) up to buildings (several metres) – serving both aesthetic as well as functional purposes. The electroplating/deposition of metallic ions (e.g. zinc, nickel, chromium) on materials requires prior coating of the material with a micrometre-thin conductive layer and has become a common method. However, this electro-based metallization suffers from two major drawbacks: firstly, the use and production of hazardous chemicals (metallic ions) and secondly, its neglect of the "materiality" of the material. "Materiality" here relates to the fact that every material, independent of its nature, is coated with a micrometre-thin conductive layer. Neglecting this materiality and the molecular interactions with a coating material can compromise its applicability. Through the exploitation of the chemical and structural (materiality-related) nature of calcium carbonate formed by micro-organisms on the one hand and those of a certain class of non-conductive polymers on the other, and their interactions, it is possible to put

▴ The combination and distribution of material in a three-dimensional space lends bone, another biomineral, its unique and superior mechanical properties.

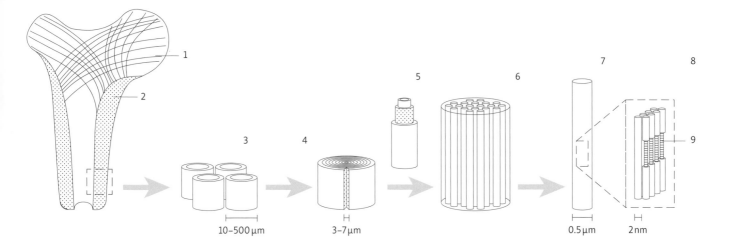

1 Macrostructure – Cancellous bone
2 Macrostructure – Cortical bone
3 Microstructure – Osteons
4 Sub-Microstructure – Haversian canal

5 Sub-Microstructure – Lamella
6 Nanostructure – Collagen fibre
7 Sub-Nanostructure – Collagen fibre

8 Sub-Nanostructure – Mineralized extracellular matrix
9 Sub-Nanostructure – Hydroxyapatite crystals

▲ **Schematic representation of the hierarchical assembly of bone.**

forward the novel and futuristic concept of growing or cultivating coatings.

Once these molecular interactions are explored, the biologically driven growth of $CaCO_3$-coatings on non-conductive polymers becomes relatively simple. The process is based on the dip-coating of polymer-based structures by immersing them into a medium containing bacteria strains (*Bacillus pasteurii*), nutrients, and a source of calcium ions. After an incubation period, the surface of the polymeric structure shows a white coverage, a 10-µm-thick layer of calcium carbonate (vaterite), which typically features a spherical morphology. The stabilization of vaterite in the form of a compact layer is attributed to organic molecules occluded into the inorganic matrix in combination with surface topology and intrinsic chemistry (molecular interactions between calcium carbonate molecules and polymer surface). Based on the chemical principle that alkaline earth-metal ions (magnesium, strontium, or barium) react with bacterial-produced carbonate ions to yield the correspondent solid carbonates, the growing material can have tailored

chemical combinations and, ultimately, new emergent properties. The approach demonstrates the controlled use of complex biological machinery towards consistent *growing coatings* on all surfaces of polymeric scaffolds across several magnitudes (from molecules to µm) and variable chemical compositions.

Dip-coating polymeric structures is an extremely versatile approach, as it permits consistent micrometre-thick coating independent of the geometry. Its sequential application leads to the biologically controlled growth of uniform, multilayered coatings similar to nacre. In addition, the dip-coating process allows variable chemical compositions so that the chemical nature of every new added layer can be tailored, yielding multilayered/multi-material biologically grown coatings, even surpassing nature's constructs like nacre.

Applications: additive manufacturing of parts
Two additional advantages associated with this process seem worth exploring: firstly, the fabrication of lightweight, high-load-bearing functions

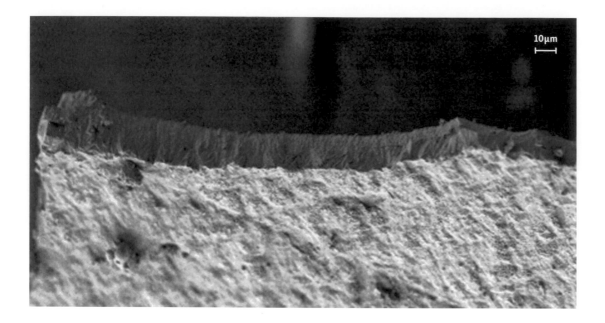

10μm

(function = material + structure) and secondly, the applicability on polymers that are currently used in additive manufacturing (3D printing). The applicability of biologically controlled growth of $CaCO_3$-coating on polymeric surfaces can be extended to biopolymers such as polylactide (PLA), unleashing its potential use in combination with additive manufacturing techniques.

Additive manufacturing of parts that are strong enough for widespread application is typically cost-prohibitive. However, a post-printing bacterial calcium carbonate treatment would efficiently strengthen 3D-printed parts for upscale applications. The process would enable freeform designs, often with complex undercuts – typically problematic for traditional manufacturing – to be produced serially and iteratively. In building applications, this approach offers the prospect of highly durable, PLA-based structures to be uniformly layered with traditional inorganic materials (limestone), which act as a substrate for architectural coatings displaying properties such as structural reinforcement.

The benefits of growing coatings in an architectural and construction context were demonstrated on a cubic module. The cube's geometric symmetry allows, contrary to nacre, a hierarchical and me-

chanical isotropy, so that all the surfaces are covered equally. The modules are assembled via a parametric, algorithmic, adaptive process along a randomly created curve. Routine structural analysis of the assemblies under compression enables to identify the modules which accumulate stress and compromise the structure's integrity. Instead of redesigning the complete structure, the modules that show accumulated stress are replaced by elements which have been repetitively dip-coated and display a controlled variable number of layers, correlating with an improved compressive strength. In this way, virtually any structure(s) can be (re)adapted to meet the needs of the building scale.

Is the concept of cultivating coatings a potential example to demonstrate that the relation between structure and materials can be unbalanced away from our current material-dependent view? Attempts to address the fundamental question of material-structure relation are not trivial and completely misleading if the currently prevailing static paradigm is followed. It is the combination of materials science and biology that holds the promise to disrupt, unbalance, and replace this static paradigm towards a new era, where materials and structures are one single dynamic entity shaped, cultivated, and grown by biological systems.

▲ Scanning electron microscopic (SEM) image of the grown 10-μm-thick layer of vaterite.

NOTES

1 Lowenstamm, H. and Weiner, S. (1989), *On Biomineralization*. Oxford: Oxford University Press.

2 Meyers, M. A., McKittrick, J., and Chen, P. Y. (2013), "Structural biological materials: critical mechanics-materials connections", *Science* 339 (6121), 773–779; Currey, J. D. (1999), "The design of mineralised hard tissues for their mechanical functions", *Journal of Experimental Biology* 202, 3285–3294.

3 Natalio, F., Corrales, T. P., Panthöfer, M., Schollmeyer, D., Lieberwirth, I., Müller, W. E., Kappl, M., Butt, H. J. and Tremel, W. (2013), "Flexible minerals: self-assembled calcite spicules with extreme bending strength", *Science* 339 (6125), 1298–1302.

4 Sarikaya, M., Aksay, I. A. (eds) (1995), *Biomimetics: Design and processing of materials*. Woodbury, NY: AIP Press.

5 Whitesides, G. M. (2015), "Bioinspiration: something for everyone", *Interface Focus* 5 (4), 20150031.

6 Ho, M. W. (2008), *The rainbow and the worm: The physics of organisms*. Singapore: World Scientific Publishing.

7 Dhami, N. K., Reddy, M. S., and Mukherjee, A. (2013), "Biomineralization of calcium carbonates and their engineered applications: a review", *Frontiers in Microbiology* 4 (314), 10–3389; Hebel, D., Wisniewska, M. H., and Heisel, F. (2015), "Building from waste", *Land Journal* Oct. 2015, 18.

BIO-INSPIRED PRODUCTS
BIOMIMICRY

"Biomimetics is the examination of nature, its models, systems, processes, and elements to emulate or take inspiration from in order to solve human problems."[1] Living organisms have evolved well-adapted structures and materials over geological time through natural selection and have achieved important engineering solutions such as self-healing, environmental exposure tolerance and resistance, hydrophobicity, self-assembly, or the harnessing of solar energy. Biomimetics has given rise to new technologies inspired by such biological solutions at macro- and nanoscales.

Products of this type operate not necessarily solely with biologically grown materials, but rather utilize any material resources. Architecture built on this basis is not an image of a biological or living system, instead it becomes a living organism itself, as the example of the BIQ House in Hamburg, Germany shows (as described by Rachel Armstrong and her colleagues, see p. 168). By integrating biological processes and their understanding into building elements (as Davide Carnelli and André Studart show in their contribution, see p. 158), materials are enabled to change their physiognomy depending on environmental influences and therefore react to their environment. Material Sciences demonstrate how even substances can become "smart" and programmed to act in certain ways or adapt to changing conditions.

If we understand a building as a living organism, it becomes an active member of our environment that is able to positively contribute to our well-being. An active building produces energy, generates oxygen, filters air and absorbs CO_2, and it can adapt to changing conditions and communicate with its neighbours to share capacities. The technical elements needed to steer and control nature-emulated processes should become part of the architectural design processes. A "vertical garden" does not necessarily need to look like a vegetable plot or a forest, as this expression might suggest (and as many designs lately tend to demonstrate). Vertical gardening could instead utilize the mechanisms of biomimetics for addressing aesthetics, climatic design, building systems, as well as oxygen and energy production.

This chapter is probably the most far-reaching concerning future developments. Speculations, visions, and first prototypes and demonstrators show the immense potential of this approach.

NOTE

1   Environment and Ecology (2016). accessed online on 14 October 2016
http://environment-ecology.com/biomimicry-bioneers/367-what-is-biomimicry.html.

Replicating Natural Design Strategies in Bio-inspired Composites

Davide Carnelli and André R. Studart

Nature has become a source of inspiration to materials scientists and engineers due to the fascinating variety of design solutions adopted by biological materials to fulfil specific structural and functional requirements.[1] The interest in looking at nature for the design of next-generation materials stems from the realization that the nano-/microstructural design found in biological systems differs markedly from that observed in synthetic materials.[2] As living organisms rely on a relatively limited choice of building blocks available in their environment, structural and functional features of natural composites rely on the design of the material at multiple lengths scales rather than on their chemical composition.[3]

Because nature offers alternative design strategies towards the fabrication of functional materials, the structure of biological materials and their unusual, environmentally benign processing routes have been investigated at multiple hierarchical levels. It is important to consider that the boundary conditions of engineering applications often differ from those found in natural environments. Therefore,

▲ Scanning electron microscopic (SEM) image of the cross section of a nacre-like material, showing biaxially aligned alumina platelets embedded in an epoxy matrix.

the key in the development of bio-inspired composites is to replicate the design principles underlying the functional response of natural structures, taking into account the relevant boundary conditions of the specific technical application rather than simply copying the chemistry and structure of biological materials per se.

The following examples showcase the identification of design principles of two functional biological materials and describe a simple processing strategy to replicate such bio-inspired design in synthetic materials which exhibit unusual toughness or self-shaping capabilities.

Toughness and self-shaping in biological structures

High toughness is often achieved in natural materials by controlling the orientation and distribution of reinforcing anisotropic particles in a softer matrix. Seashells are primary examples of structural biological materials whose reinforcement architecture is designed to achieve a remarkable enhancement of mechanical properties in spite of using

▲ **View of the inner nacreous layer of a red abalone shell.**

mechanically poor building blocks.[4] The secret of the seashell's mechanical performance is its highly oriented structure of stiff platelets in a tight regular "brick-mortar" arrangement within the bottom layer, called nacre. This micro-structural design allows for the development of an effective combination of toughening mechanisms which include delocalization of damage on large areas of the material, crack deflection, and bridging of the faces of the crack.[5] This results in toughness values that are at least two orders of magnitude higher than that of its main constituents.

Precise control of the orientation of reinforcing elements is also a distinguished structural feature of plant seedpods. In this case, orientation control in bilayers is used to achieve impressive macroscopic shape changes that enable the release of seeds when the relative humidity varies.[6] One example is represented by the orchid tree seedpod, which undergoes structural twisting upon changes in the water content within the pod. This change in the geometrical configuration is possible due to the

ordered structure of cellulose microfibrils, which are embedded in a water-absorbing hemicellulose matrix at the plant cell walls of individual layers.[7] The orientation of cellulose fibrils along different directions in the two layers causes local differential swelling or shrinkage of adjacent layers during hydration. This generates bending and stretching deformations which can only be relaxed by changing the shape of the seedpod into well-defined macroscopic geometries.

Mimicking biological design principles by the magnetic alignment technology

Mimicking in synthetic systems the biological design principles described above requires tools that allow for controlling the material's reinforcement design at the microscopic level. In conventional man-made composites, this control represents a challenging technological task. This manufacturing limitation has been overcome by using external magnetic fields to remotely align reinforcing particles suspended in a liquid, which is later removed or converted into a solid phase.[8] Similar to metallic

▲ Scanning electron microscopic (SEM) image of the bilayered structure encountered in seashells, consisting of nacreous (bottom) and prismatic (middle) layers.

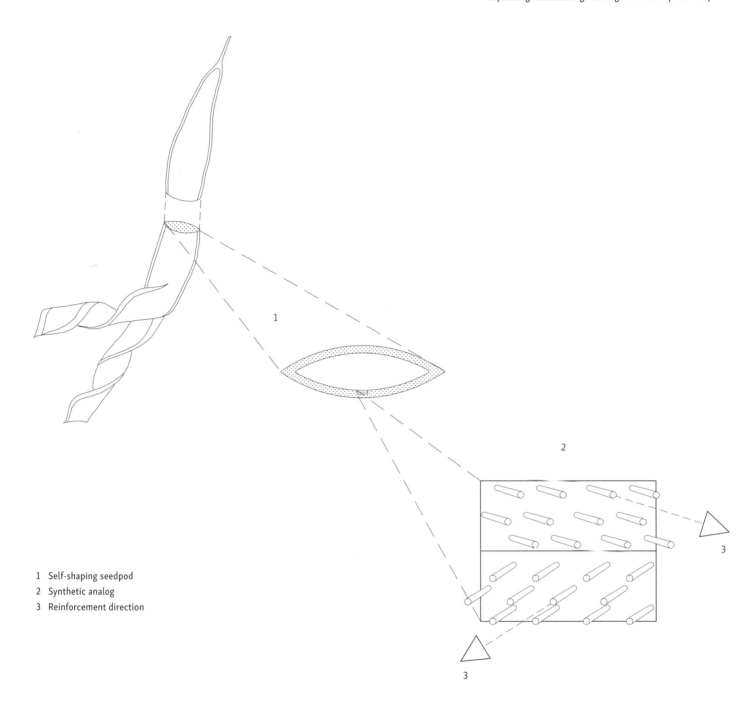

1 Self-shaping seedpod
2 Synthetic analog
3 Reinforcement direction

▲ Cartoon schematic indicating the
predominant orientation of cellulose
microfibrils within the orchid tree
seedpod, and the synthetic analogue.

◄ Open and closed state of an orchid
tree seedpod when dry or wet, respectively.

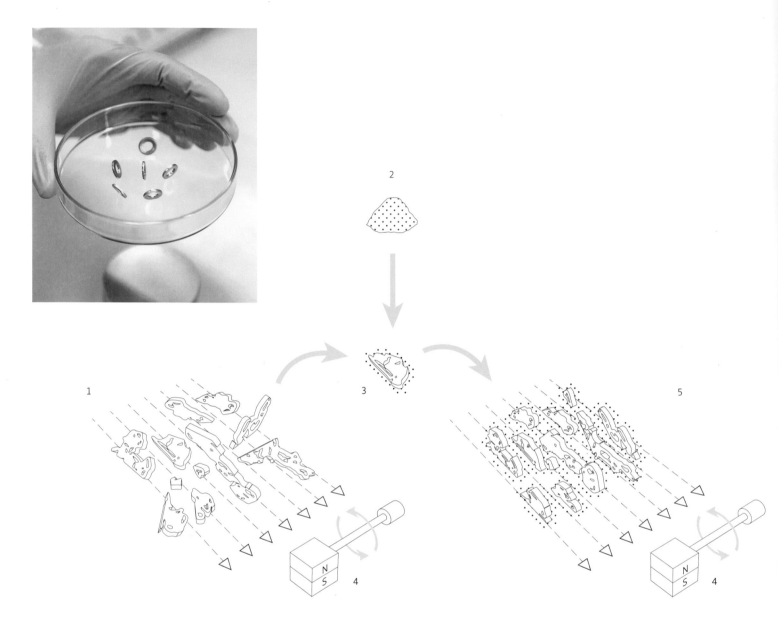

1 Unaligned platelets
2 Magnetic particles
3 Platelet coated with magnetic particles
4 Rotating magnet
5 Coated platelets self-aligning to magnetic force field

washers that align themselves along the direction of the magnetic field when approaching a permanent magnet, reinforcing stiff particles that are 1,000 times smaller in size can also be made magnetically responsive by coating them with superparamagnetic iron oxide nanoparticles. By employing this technique, it is possible to deliberately control the alignment of microparticles in one or two dimensions with a static or rotating magnetic field.

When the procedure described above is applied to align micrometre-sized, stiff platelets in an organic matrix, highly oriented structures that resemble the nacreous layer of seashells can be obtained.[9] The use of stiff inorganic platelets as reinforcement in bioinspired composites ensures a high stiffness and high bending strength. The highly aligned platelet structure also allows for the implementation of toughening mechanisms similar to those found in nacre.[10] One example of such mechanisms

▲ Out-of-plane, unidirectional alignment of magnetic washers by subjecting them to a magnetic field arising from a permanent magnet.

▼ Procedure of aligning stiff, non-magnetic reinforcing microparticles after electrostatic adsorption of iron oxide nanoparticles on their surface.

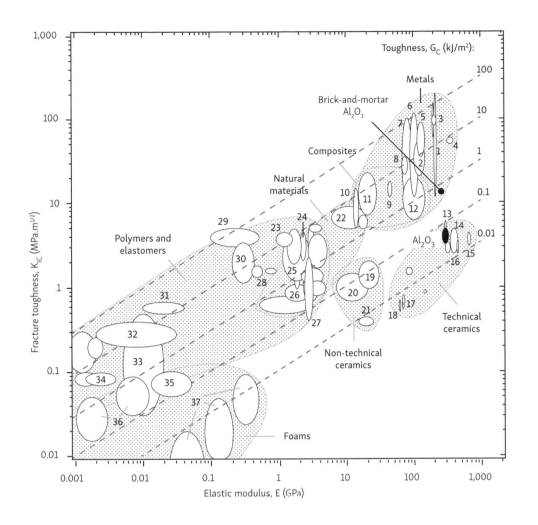

Toughness, G_C (kJ/m²):

Metals

Brick-and-mortar Al_2O_3

Composites

Natural materials

Polymers and elastomers

Technical ceramics

Al_2O_3

Non-technical ceramics

Foams

1 Steels
2 Cast irons
3 Ni alloys
4 W alloys
5 Cu alloys
6 Ti alloys
7 Zn alloys
8 Al alloys
9 Mg alloys
10 Lead alloys
11 GFRP
12 CFRP
13 Si3N4
14 SiC
15 WC
16 B4C
17 Silica glass
18 Soda glass
19 Brick
20 Stone
21 Concrete

22 Wood
23 PP
24 PC
25 ABS
26 PS
27 Epoxies
28 PTFE
29 Leather
30 Ionomers
31 EVA
32 Polyurethane
33 Silicone elastomers
34 Butyl rubber
35 Cork
36 Rigid polymer foams
37 Flexible polymer foams
◂

Fracture toughness, K_{IC} (MPa.m$^{1/2}$)

Elastic modulus, E (GPa)

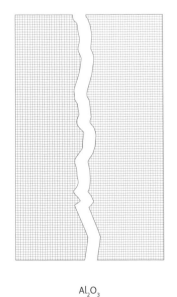

Al_2O_3

Al_2O_3/PMMA

▴ **Fracture toughness K_{IC} of a wide range of man-made structural materials as a function of their elastic modulus E. Indicated is the high toughness of a bio-inspired brick-and-mortar Al_2O_3 ceramic assembled from inherently weaker alumina building blocks.**

◂ **Straight path of a crack as it propagates through bulk, unstructured Al_2O_3 (left). Tortuous crack path obtained for a synthetic nacre-like material containing an interconnected brick-and-mortar structure of Al_2O_3 platelets in epoxy (right).**

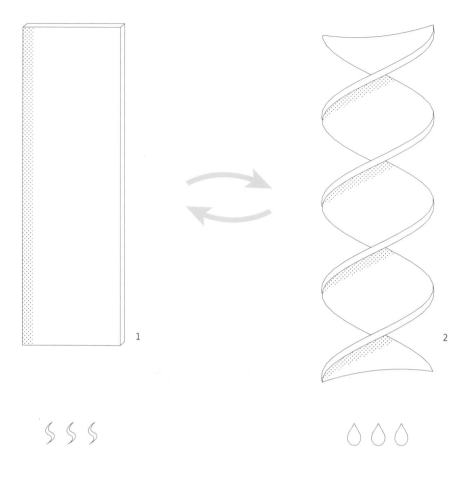

1 Dry biocomposite
2 Moist biocomposite

in a bio-inspired synthetic composite is the tortuous path that a main crack experiences at the surface of highly aligned alumina platelets as it propagates perpendicular to the alignment direction. The toughness of nacre-inspired alumina-based composites with an internal brick-mortar architecture can reach values that are three to six times higher than that of pure alumina.

Besides the nacreous layer of mollusc shells, the magnetic alignment technology has also been utilized to emulate the self-shaping functionalities of plant seedpods.[11] Self-shaping synthetic seedpods with chiral twisting functionalities can be triggered by changes in the water content of the composite bilayer. Self-shaping in the form of twisting is achieved thanks to the magnetically programmed orientation of reinforcing alumina platelets at angles of $+\pi/4$ $(-\pi/4)$ within a swellable polymer matrix.

Mimicking the biological layer-by-layer process: printed, self-shaping functional objects

In addition to the organization of reinforcing elements in deliberate directions, biological materials also exploit structural heterogeneities and gradients through a layer-by-layer assembly process controlled by the living organism. For instance, in biomineralization, the mineral phase is sequentially added into a pre-structured organic scaffold, allowing for site-specific control over the chemical composition and reinforcement orientation. The layer-by-layer nature of the biological assembly strategy has been recently exploited using additive manufacturing techniques to create a wide range of heterogeneous, bio-inspired materials.[12] As an example, a 3D printing approach known as direct ink writing was recently combined with the magnetic alignment technology discussed above to create self-shaping functional objects with

▲ Self-shaping functionality of synthetic materials with bilayer reinforcement architectures inspired by orchid tree seedpods.

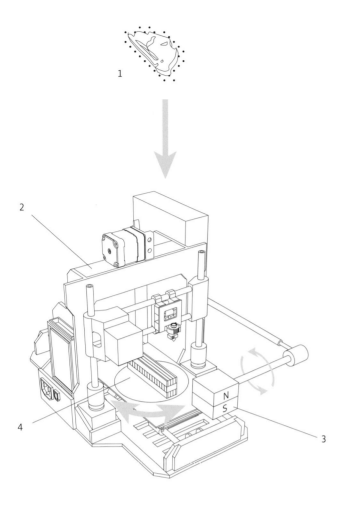

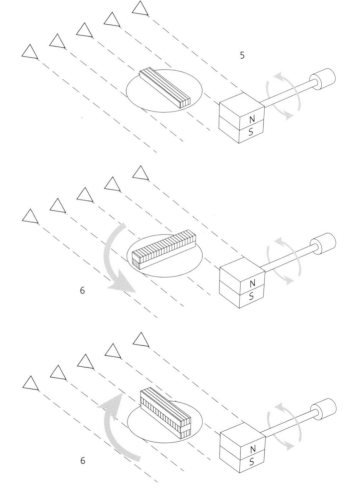

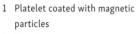

1   Platelet coated with magnetic
    particles

2   3D printer

3   Rotating magnet

4   Product during printing process

5   Direction of magnet influencing
    platelet orientation during printing

6   Turning printbed can create print
    layers with changing platelet orien-
    tations.

▴ **The layer-by-layer additive approach of 3D printing technology can
be utilized to vary material properties
through directional changes in the
magnetic force field in the printing of
biologically inspired materials.**

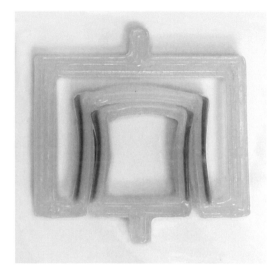

site-specific control of chemical composition and particle orientation.[13] Such level of control is exploited to induce differential expansion or contraction of the walls of the printed object to enable the autonomous fastening of a key-lock soft joint that can carry a rigid object much heavier than its own weight.

These examples illustrate how man-made technologies have reached unprecedented freedom in structural design from the nano- to the micro- and macroscopic scales, opening a path towards the fabrication of structures with thus far unforeseen functionalities for a broad range of applications. Using design principles that have been probed through hundreds of millions of years of evolution in nature, such manufacturing technologies offer a powerful platform for the future design and fabrication of more durable and sustainable functional materials and structural elements.

▲ Self-shaping soft objects with complementary key-lock geometries fabricated by a DIW approach called multimaterial magnetically-assisted 3D printing.

NOTES

1 Studart, A. R. (2012), "Towards high-performance bio-inspired composites", *Advanced Materials* 24 (37), 5024–5044; Ritchie, R. O. (2011), "The conflicts between strength and toughness", *Nature Materials* 10 (11), 817–822.

2 Bonderer, L. J., Studart, A. R., and Gauckler, L. J. (2008), "Bioinspired design and assembly of platelet reinforced polymer films, *Science* 319 (5866), 1069–1073; Erb, R. M., Libanori, R., Rothfuchs, N., Studart, A. R. (2012), "Composites reinforced in three dimensions by using low magnetic fields", *Science* 335 (6065) ,199–204; Munch, E., Launey, M. E., Alsem, D. H., Saiz, E., Tomsia, A. P., Ritchie, R. O. (2008), "Tough, bio-inspired hybrid materials", *Science* 322 (5907) 1516–1520; Bouville, F., Maire, E., Meille, S., Van de Moortele, B., Stevenson, A. J., Deville, S. (2014), "Strong, tough and stiff bioinspired ceramics from brittle constituents", *Nature Materials* 13 (5), 508–514; Hull, D., Clyne, T. W. (1996), *Toughness of composites: An Introduction to Composite Materials*. Cambridge: Cambridge University Press.

3 Meyers, M. A., Chen, P.-Y., Lin A.Y.-M., and Seki, Y. (2008), "Biological materials: Structure and mechanical properties", *Progress in Materials Science* 53 (1), 1–206; Fratzl, P. and Weinkamer, R. (2007), "Nature's hierarchical materials, *Progress in Materials Science*, 52 (8), 1263–1334.

4 Sun, J. and Bhushan, B. (2012), "Hierarchical structure and mechanical properties of nacre: a review", *RSC Advances* 2 (20), 7617–7632; Currey, J. D. (1977), "Mechanical properties of mother of pearl in tension", *Proceedings of the Royal Society B*, 196, 443–463; Barthelat, F., Li, C.-M., Comi, C., and Espinosa, H. D. (2006), "Mechanical properties of nacre constituents and their impact on mechanical performance", *Journal of Materials Research* 21 (08), 1977–1986.

5 Launey, M. E. and Ritchie, R. O. (2009), "On the fracture toughness of advanced materials", Advanced Materials 21 (20), 2103–2110; Xia, S., Wang, Z., Chen, H., Fu, W., Wang, J., Li, Z., and Jiang, L. (2015), "Nanoasperity: Structure origin of nacre-inspired nanocomposites", *ACS Nano* 9 (2), 2167–2172; Wang, R. Z., Suo, Z., Evans, A. G., Yao, N., and Aksay, I. A. (2001), "Deformation mechanisms in nacre", *Journal of Materials Research* 16 (09), 2485–2493.

6 Elbaum, R., Zaltzman, L., Burgert, I., and Fratzl, P. (2007), "The role of wheat awns in the seed dispersal unit", *Science* 316 (5826), 884; Burgert, I., and Fratzl, P. (2009), "Actuation systems in plants as prototypes for bioinspired devices", *Philosophical Transactions of the Royal Society A* 367 (1893), 1541; Harrington, M. J., Razghandi, K., Ditsch, F., Guiducci, L., Rueggeberg, M., Dunlop, J. W. C., Fratzl, P., Neinhuis, C., and Burgert, I. (2011), "Origami-like unfolding of hydro-actuated ice plant seed capsules", *Nature Communications* 2, 337.

7 Emons, A. M. C. and Mulder, B. M. (2000), "How the deposition of cellulose microfibrils builds cell wall architecture", *Trends in Plant Science* 5 (1), 35–40; Burgert, I. and Fratzl, P. (2009), "Plants control the properties and actuation of their organs through the orientation of cellulose fibrils in their cell walls", *Integrative and Comparative Biolology* 49 (1), 69–79.

8 Erb, R. M., Libanori, R., Rothfuchs, N., and Studart, A. R. (2012), "Composites reinforced in three dimensions by using low magnetic fields", *Science* 335 (6065), 199–204.

9 Erb, R. M., Libanori, R., Rothfuchs, N., and Studart, A. R. (2012), "Composites reinforced in three dimensions by using low magnetic fields", *Science* 335 (6065), 199–204; Libanori, R., Erb, R. M., and Studart, A. R. (2013), "Mechanics of Platelet-Reinforced Composites Assembled Using Mechanical and Magnetic Stimuli", *ACS Applied Materials & Interfaces* 5 (21), 10794–10805.

10 Grossman, M., Bouville, F., Erni, F., Masania, K., R. Libanori, R., and Studart, A. R. (2017), "Mineral nano-interconnectivity stiffens and toughens nacre-like composite materials", *Advanced Materials* 29(8), 1605039.

11 Erb, R. M., Sander, J. S., Grisch, R., and Studart, A. R. (2013), "Self-shaping composites with programmable bioinspired microstructures", *Nature Communications* 4, 1712; Bargardi, F., Le Ferrand, H., Libanori, R., and Studart, A. R. (2016), "Bioinspired self-shaping ceramics", *Nature Communications* 7, 13912.

12 Studart, A. R., (2016), "Additive manufacturing of biologically-inspired materials", *Chemical Society Reviews* 45 (2), 359–376.

13 Kokkinis, D., Schaffner, M., and Studart, A. R. (2015), "Multimaterial magnetically assisted 3D printing of composite materials", *Nature Communications* 6, 8643.

Living Architecture (LIAR): Metabolically Engineered Building Units[1]

Rachel Armstrong, Andrew Adamatzky,
Gary S. Caldwell, Simone Ferracina,
Ioannis Ieropoulos, Gimi Rimbu, José Luis García,
Juan Nogales, Barbara Imhof, Davide de Lucrezia,
Neil Phillips, and Martin M. Hanczyc

◄ Bioreaction within Arup's state-of-the-art BIQ (Intelligent Building) House in Hamburg, 2013. The algae serve as a form of micro-agriculture.

1 Bacteria anaerobically oxidize a range of organic substrates, respiring electrons (as a by-product) and realizing cations such as protons.

2 Electrons then flow via an external circuit forming the electrical current, whereas protons diffuse through the proton-exchange-membarn (PEM) to recombine at the cathode and react with oxygen to produce water.

3 Algae in the MFC cathodic environment photosynthesize and produce oxygen as a by-product.

4 Anode

5 Cathode

▲ Scheme of the intended final configuration of the microbial fuel cell (MFC) with a photocathode used as a modular building block wherein bacteria generate bioelectricity through the oxidation of organic wastes that can be coupled with renewable biomass.

Buildings are often described in terms that infer they possess "life". Architectural design, in the same spirit, frequently explores ideas about the living world, the body and nature – through various approaches to the construction and choreography of space. Yet, most buildings are constructed according to industrial paradigms based on inert materials, or materials that were formerly, but are no longer, "lively" such as wood, bricks, and soils, materials that are processed into inert forms obedient to top-down design processes.

We are currently entering an ecological era in which we are increasingly viewing reality as a hyper-complex and interconnected, open system with constant fluxes of energy, matter, and information.

This shift has become an everyday reality with, for example, the advent of the Internet, which has allowed us to transgress previously irreducible divides of geography, culture, politics, matter, and identity. Yet the ecological era does not concern the greening of things, or simply substituting an object-centred view of reality with a process-oriented one. Rather, it requires us to develop technical systems that speak to the principles of design and engineering with living systems. The innate properties of life-like systems exceed the expectations of traditional approaches of construction and invite the development of an ecological approach to architectural practice that invokes the cultivation of building materials. This approach is predicated on the dynamic adaptability and variation

▲ **Façade detail of the BIQ House. Energy savings were made within the building through the solar-thermal effect of algae biomass.**

▲ The BIQ House grows and harvests
algae biomass that reduces both
energy demand and CO_2 emissions
compared to a common building
system. It was part of the International Building Exhibition in Hamburg,
2007–2013.

implicit in living systems, and it shifts away from the traditional view of architecture as a static, form-giving subject that is centred on making a building, but rather moves towards developing alternative ways of choreographing matter in relation to populations and sites.

Life's systems as technological intervention

In an age of biotechnology, new ideas, materials, and technologies enable us to consider life itself as a technology and to not only manipulate life's fabrics, but also use life's systems themselves as the technical intervention. Transposing the technical competencies of biological systems from the nano-microscale to the architectural-human scale is challenging. The issues of scale can be addressed,

in an architectural context, by seeking technological convergence and coordinating metabolic materials or agents in space.

In 2013, the international engineering company Arup built the first publicly accessible algae façade for the BIQ House in Hamburg, Germany, which served as a form of micro-agriculture. In addition, energy savings were made within the building through the solar-thermal effect of the algae biomass.[2] More recently, the Senseable City Laboratory at the Massachusetts Institute of Technology in Cambridge, Massachusetts, and the ecoLogicStudio in London produced an algaeponics design for the 2015 Milan Expo.[3] These architectures function as cellular gardens that are farmed but do not engage

▴ **Algae biomass is one measurable indicator of a living architecture, which is capable of many products.**

with the principles of directly designing with life. Even so, such concepts are of interest and widely entertained in architectural design from Terraform One's In Vitro Meat Habitat, which proposes to culture an entire building from laboratory-grown pig cells,[4] to bioluminescent trees created with the help of genes (GFP – Green Fluorescent Protein) originally from bioluminescent jellyfish, proposing to supply off-grid, low-level lighting in the Genetic Barcelona Project of the Genetic Architectures Research Group at the Universidad Internacional de Cataluña.[5]

Beyond biomimicry

Biomimicry proposes to copy or be inspired by nature to produce more efficient solutions.[6] Be-

yond biomimicry, a field entitled biomimetics aims to not only be inspired by nature, but translate biological principles into mechanisms, technologies, or other technical systems. It remains the question whether there is even a step beyond biomimicry? An answer could arise from the staggering advances of the past decade in the life sciences, particularly in metabolic engineering, where "the realm of the born – all that is nature – and the realm of the made – all that is humanly constructed – are becoming one".[7] Potentially, these synergies could be harnessed heralding a new age in the design and engineering of living things.[8] It is this disruptive vision that the Living Architecture (LIAR) concept seeks to explore.[9] LIAR integrates a suite of established (algaeponics, microbial fuel cells,

▲ BBiC's microbial fuel cell (MFC) in operation, as exhibited at the Biennale, 15th International Architecture Exhibition, in Venice in 2016.[10]

unconventional computing, and robotics) and yet-to-be-established technologies and processes to produce a unique modular computational metabolism construct, a construct that is to illustrate the scope for next-generation living systems at an architectural scale.

Living Architecture (LIAR) for the built environment

Living Architecture (LIAR) is conceived to produce standardized components (architectural, robotic, metabolic, and control software) as an operational unit of construction for an ecological era. Cultivated building materials offer potentials to adapt the living environment to meet the needs and desires of the occupants. Such a kind of functional architecture has not yet been built, but a consideration of its technical feasibility begins to outline some of the challenges. We anticipate such a structure to take the shape of bioreactor walls that can be installed in open environments within modern buildings. LIAR is based on the operational principles of microbial communities activating both microbial fuel cells (MFCs) and photo bioreactors as new environments.

These biological components are metabolically engineered to deliver specified biochemical and systematic functions. LIAR is conceived to become an integral component of human dwellings capable of extracting valuable resources from light, wastewater, and flue gas/exhaust air. It is to polish wastewater, generate oxygen, and produce useable biomass, fertilizer, and electrical power. As a self-learning, self-governing system it will adapt to local conditions, thereby maintaining high system efficiencies by selectively manipulating consortia performance.

A freestanding partition is conceived as a LIAR prototype, consisting of modular bioreactor units designed for incorporation into modern spaces with traditional utilities. To this purpose, the peripheral physical dimensions of the bioreactors have been defined as multiples of standard building blocks (e.g. bricks or concrete blocks). This will allow them to be effectively nested within walls if not left on display in open spaces, while maintaining both structural integrity and providing thermal mass. In the longer term, LIAR is aimed at applications within the built environment to address global-scale challenges relevant to urban sustainability, resource management, and the circular economy.[11]

Impacts

The current model proposed for LIAR has potential commercial relevance in the development of home appliances, e.g. washing machine technology and water purifiers for kitchens and bathrooms. In this way, living spaces are to become programmable ecosystem modules that can be customized to suit the resources and needs of a community. Commercially speaking, LIAR would not compete with industrial-scale operations of production, but aims to add value to the health of environments, recycling scarce resources and supporting products with reusable or tradable value.

LIAR has the potential for far-reaching and transferable impacts on the performance of our living spaces and cities. It implies a new understanding of sustainability[12] that goes beyond resource conservation and incorporates the metabolic design of living spaces. Achieving this conceptual and technical breakthrough may endow our cities with robustness and resilience to the impacts of climate change while enabling inhabitants to live humanely and even profitably in the highly resource-constrained and competitive circumstances of the future.

NOTES

1 The project has received funding from the European Union's Horizon 2020 Research and Innovation Programme under Grant Agreement no. 686585.

2 Steadman, I. (2013), "Hamburg unveils world's first algae-powered building", *Wired* [website], http://www.wired.com/design/2013/04/algae-powered-building/, last accessed on 14 July 2015.

3 Medina, S. (2014), EcoLogicStudio, *Metropolis Magazine* [website], http://www.metropolismag.com/October-2014/EcoLogic-Studio/., last accessed 14 July 2015.

4 Armstrong, R. and Spiller, N. (2011), "Synthetic biology: Living quarters", *Nature 467*, 916–918.

5 Myers, W. and Antonelli, P. (2013), *Bio Design: Nature, Science, Creativity*. London: Thames & Hudson/New York: MOMA, 68–69.

6 Beynus, J. (1997), *Biomimicry: Innovation inspired by Nature*. New York: William Morrow.

7 Kelly, K. (1994), *Out of Control: The Rise of Neo-Biological Civilization*. Boston: Addison-Wesley, 1994.

8 Venter, J. Craig (2013). *Life at the speed of light: From the Double Helix to the Dawn of Digital Life*. London: Little, Brown.

9 Living Architecture (2016) (website), http://livingarchitecture-h2020.eu, last accessed 9 July 2016.

10 Bristol BioEnergy Centre http://www.brl.ac.uk/researchthemes/bioenergyself-sustaining.aspx

11 Ellen MacArthur Foundation (no date), *Circular Economy* (website), http://www.ellenmacarthurfoundation.org, last accessed 29 October 2015.

12 World Commission on Environment and Development (1987), *Our Common Future*. Report of the World Commission on Environment and Development, published as Annex to General Assembly document A/42/427, http://www.un-documents.net/our-common-future.pdf, last accessed 14 July 2015.

About the Authors and Contributors

Dirk E. Hebel is Professor of Sustainable Construction at the Karlsruhe Institute of Technology (KIT) in Germany as well as the Future Cities Laboratory (FCL) in Singapore. He previously held the position of Assistant Professor of Architecture and Construction at the Department of Architecture of ETH Zürich in Switzerland; he was the Founding Scientific Director of the Ethiopian Institute of Architecture, Building Construction and City Development in Addis Ababa. Between 2002 and 2009, he taught at the Department of Architecture at ETH Zürich, held a guest professorship at Syracuse University and taught as a guest lecturer at Princeton University, USA. Work resulting from his teaching and research has been published in numerous books and academic journals, lately *SUDU – The Sustainable Urban Dwelling Unit* (Ruby Press, 2015) and *Cities of Change – Addis Ababa* (with Marc Angélil, second and revised edition, Birkhäuser, 2016). His research concentrates on a metabolic understanding of resources and investigates alternative building materials and construction techniques and their applications in developed as well as developing territories.

Felix Heisel is an architect currently working as Head of Research and PhD candidate at the Professorship of Sustainable Construction Dirk E. Hebel at the Karlsruhe Institute of Technology (KIT) in Germany as well as the Future Cities Laboratory (FCL) in Singapore. He previously held the position of Research Coordinator at the Professorship of Architecture and Construction at ETH Zürich. He has taught and lectured at the ETH Zürich and the Future Cities Laboratory in Singapore, as well as the Ethiopian Institute of

Architecture, Building Construction and City Development in Addis Ababa, the Berlage Institute in Rotterdam, and the Berlin University of the Arts. His teaching and research has been published in books and academic journals including *Building from Waste* (with Dirk E. Hebel and Marta H. Wisniewska, Birkhäuser, 2014). His interest in informal processes led him to establish the documentary movie series _Spaces in 2011, which is published online and as the book *Lessons of Informality* (with Bisrat Kifle, Birkhäuser, 2016). His research work is focused on the phenomena of informality in connection and dependency on the establishment of (alternative) building materials in developing territories.

Dieter Läpple is Professor Emeritus of International Urban Studies at the HafenCity University Hamburg. For many years he directed the Institute for Urban Economics at the Hamburg University of Technology. He was Guest Professor at the Aix-Marseille University and the Institut d'Études Politiques de Paris, as well as advisory member of the Urban Age Programme at the London School of Economics and Senior Fellow of the Brookings Institution in Washington. As member of the Board of Trustees, he helped shape the International Building Exhibition (IBA) Hamburg.

Ákos Moravánszky is Professor Emeritus of Theory of Architecture at the Institute for the History and Theory of Architecture gta at ETH Zürich. He has served on several editorial boards including the architectural journals *Werk, Bauen+Wohnen* and *tec21*. He is advisor to the Getty Grant Program in Los Angeles. A main area of his research and publication activities is the iconology of building materials and constructions.

Nikita Aigner is a wood scientist with a main research focus on the development of novel lignocellulose-based materials. **Philipp Müller** is a carpenter and civil engineer by training and has worked world-wide in the structural design of timber and earth buildings. At the time of writing, both are researchers at the Professorship of Dirk E. Hebel at the Future Cities Laboratory (FCL) in Singapore.

Christof Ziegert is a bricklayer and civil engineer, as well as partner of Ziegert I Roswag I Seiler Architekten Ingenieure (ZRS) in Berlin. He is board member of the German earth building organization Dachverband Lehm e.V., chairman of the earth building standards committee at the German Standards Institute (DIN), and Honorary Professor of Earth Building at Potsdam University of Applied Sciences. Currently, he acts as board member and chairman of ICOMOS-ISCEAH. **Jasmine Alia Blaschek** is an architect currently employed as assistant at ZRS in Berlin, Germany.

Aliriza Javadian is a civil engineer and post-doc researcher. He finished his PhD research on a novel bamboo composite material at the Future Cities Laboratory (FCL) in Singapore. **Simon Lee** is a mechanical engineer and focuses on the conceptual development and structural analysis of a novel manufacturing method for composite materials. **Karsten Schlesier** is a civil engineer and PhD candidate focusing on the design and evaluation of alternative construction systems with an emphasis

on cultivated building materials. At the time of writing, all three are working at the Professorship of Sustainable Construction at the Karlsruhe Institute of Technology (KIT) in Germany as well as the Future Cities Laboratory (FCL) in Singapore.

Felix Böck graduated as Wood Technology & Industrial Engineer from Rosenheim University of Applied Sciences and is currently at work on his PhD at the University of British Columbia in Vancouver. He focuses on technologies for processing bamboo for the development of bamboo-wood hybrid elements, with the goal to develop products which lead to new business ideas and opportunities in the rapidly growing economies of developing territories.

Peter Edwards and Yvonne Edwards-Widmer are plant ecologists. Peter is a Professor at ETH Zürich and director of the Singapore-ETH Centre. His current work concerns the ecosystem services provided by trees in tropical cities. Yvonne conducted her PhD research on the ecological role of bamboos in old-growth oak forests in Costa Rica. She has collaborated with Costa Rican institutions in promoting knowledge about the sustainable use of bamboo.

Pablo van der Lugt finished his PhD research about the environmental impact of industrial bamboo materials at Delft University of Technology in 2008, followed by various ambassador roles in the green building industry. At the time of writing he is Head of Sustainability at MOSO International and Accsys Technologies as well as guest lecturer for bio-based materials at Delft University of Technology.

Frauke Bunzel studied chemistry and received her doctoral degree at the University of Braunschweig in the area of macromolecular chemistry. **Nina Ritter** studied forest science and forest ecology at the University of Göttingen, her doctorate focused on lightweight wood-based materials. Both are researchers at the Fraunhofer Institute of Wood Research, Wilhelm-Klauditz-Institut WKI, working among others on the development of wood foam.

Leon van Paassen obtained an MSc and PhD and worked for several years as Assistant Professor at Delft University of Technology, investigating bio-based processes for geotechnical engineering applications. At the time of writing he is Associate Professor Biogeotechnics at Arizona State University. **Yask Kulshreshtha** obtained his MSc and worked as a researcher at Delft University of Technology. He is the inventor of CoRncrete, a construction material using corn starch as a cement replacing binder.

Hanaa Dahy is an architect and a Junior Professor leading the Bio-based Materials and Materials Cycles in Architecture (BioMat) department at the Institute for Building Structures and Structural Design (ITKE) of the Faculty of Architecture and Urban Planning at the University of Stuttgart. **Jan Knippers** is a civil engineer and a Professor, leading the ITKE. They both cooperate in material developments and their diverse applications in architecture.

Peter Axegård has 20 years of experience in bio-refining with a focus on industrial production of new bio-based products from cellulose, lignin, and hemicelluloses in or connected to pulp mills. **Per Tomani** has been active for 15 years in lignin separation from kraft pulp mills and lignin valorization for bio-based lignin products on a lab to industrial scale. Both are currently employed at INNVENTIA AB, as Vice President, Bioeconomy Strategy and as Business Development Manager, Biorefinery and Biobased Materials.

Phil Ross is an artist and a bioengineering scholar. His work has been exhibited at the Venice Biennale of Architecture, the Museum of Modern Art in New York and the Museum of Jurassic Technology in California. At Stanford University, he works on developing an internet of biological things. He is also co-founder and CTO of MycoWorks, a startup that turns mushroom mycelium into building bricks and leather.

Henk M. Jonkers studied Marine Biology at Groningen University, where he also obtained his PhD in Marine Microbiology. After working for seven years as research scientist at the Max Planck Institute for Marine Microbiology in Bremen, he joined the Faculty of Civil Engineering of Delft University of Technology as Associate Professor in the Materials & Environment section. His interests and current research projects include the development of bio-based construction and self-healing materials.

Filipe Natalio is a chemist with a broad range of interests, currently working at the Kimmel Center for Archaeological Science at the Weizmann Institute of Science. Filipe's research has always aimed at the interface between different natural sciences disciplines. His most recent interest lies in archaeological sciences in order to improve information exchangeability between chemical science and human behaviour.

Davide Carnelli is a mechanical engineer with MSc and PhD degrees from Politecnico di Milano. After research appointments at the MIT in Cambridge and EPFL Lausanne, he is currently Postdoc Researcher at the Complex Materials Group of ETH Zürich. **André R. Studart** is Associate Professor for Complex Materials at ETH Zürich. He is a materials scientist with BSc and PhD degrees from the Federal University of São Carlos and post-doctoral experience at ETH Zürich and Harvard University.

Rachel Armstrong is Professor of Experimental Architecture at the University of Newcastle. She is a 2010 Senior TED Fellow and sustainability innovator who investigates a new approach to building materials called "living architecture", which suggests our buildings may share some of the properties of living systems. Living Architecture (LIAR) includes experts from the universities of Newcastle (England), the West of England (England) and Trento (Italy), in collaboration with the Spanish National Research Council (Spain), LIQUIFER Systems Group (Austria) and EXPLORA (Italy).

Acknowledgments

We would like to express our sincere gratitude to all authors and contributors to this book for their involvement and dedicated support. Further, we would like to offer our thanks to the Singapore-ETH Centre (SEC) for Global Environmental Sustainability and the Future Cities Laboratory (FCL), particularly SEC Managing Director Dr Remo Burkhard, who made this publication possible, SEC Director Prof. Dr Peter Edwards, Founding SEC Director Prof. Dr Gerhard Schmitt, FCL Scientific Director Prof. Dr Stephen Cairns, and FCL Program Leader Prof. Kees Christiaanse, as well as the Department of Architecture of ETH Zürich, Prof. Annette Spiro. We are extremely thankful for the continuous support of the President of ETH Zürich, Prof. Dr Lino Guzzella, former President Prof. Dr Ralph Eichler, Rector Prof. Dr Sarah Springman, and Vice President of Research, Prof. Dr Detlef Günther. We would like to extend our gratitude also to our own natural fibre research team in Singapore and Zurich, Dr Alireza Javadian, Simon Lee, Philipp Müller, Karsten Schlesier, Tobias Eberwein, Mateusz Wielopolski, and Dr Nikita Aigner and to our research partners at Empa Dübendorf, Dr Peter Richner, Dr Tanja Zimmermann, Dr Sébastien Josset, and Walter Risi. Our thanks also go to REHAU, Nils Wagner, Dr Dragan Griebel, and Dr Mateusz Wielopolski. We further thankfully acknowledge the financial support of our bamboo composite research by the Swiss Commission for Technology and Innovation (CTI project no 17198.1 PFIW-IW), the Singapore-MIT Alliance for Research and Technology (SMART Innovation Grant ENG13001), as well as the National Research Foundation in Singapore. We would like to extend our gratitude to Miriam Bussmann, the graphic designer of this book, and to Sophie Nash for her outstanding graphic representations. Our special thanks go to the editor for the publisher, Andreas Müller, who made this book possible through his great dedication and creative input.

Illustration Credits

adidas Group 21/1–2

Archivist/Alamy Stock Photo 35

Armstrong, Rachel 172, 173

BIOMAT Research group/ITKE University of Stuttgart (Hanaa Dahy) 116, 121, 122

Birkhäuser/DeGruyter, Dirk E. Hebel, Marta H. Wisniewska and Felix Heisel 15

Böck, Felix 72, 73/3, 74; (illustrated by Sophie Nash) 75; 77

Bogdan Ionescu/Shutterstock.com 159

Braun, Zooey 16/2

Chinnasot, Teerasak/shutterstock.com 137

Colt International, Arup, SSC GmbH 170, 171

Deville, Sylvain 150/1

Diewald, Christoph 161/2

Diglas, Andrea ITA/Arch-Tec-Lab AG 36/1, 38

ERNE AG Holzbau (illustrated by Guido Köhler) 36/3

Eye of Science/Science Photo Library 160

Fraunhofer WKI (A. Gohla) 103/1–2

Fraunhofer WKI (M. Lingnau) 106

Gramazio Kohler Research ETH Zürich 36/2, 37

Graves, Chris 18

Hammer, Manfred 118, 119/2, 120

Hancock, Thomas 27, 28

Herman Miller 16/1

HewSaw Machines Inc. 34/2

Hidalgo-Lopez, Oscar (2003). Bamboo – The Gift of the Gods 54/1, 59/2

IBA Hamburg GmbH (Kai Dietrich) 22/1

IBA Hamburg GmbH (Martin Kunze) 22/2

iStock.com/alanphillips 138/1

iStock.com/AndreasReh 17/3

iStock.com/bernsmann 45/1

iStock.com/Cassis 138/2

iStock.com/Diosmirnov 32/2

iStock.com/ET1972 32/1

iStock.com/herraez 98–99

iStock.com/hippostudio 54/2

iStock.com/jarun011 132, 133

iStock.com/KarenHBlack 156, 157

iStock.com/Lizalica 17/1

iStock.com/luismmolina 142

iStock.com/nanoqfu 84

iStock.com/SoumenNath 52

iStock.com/Udomsook 136

Jonkers, Henk M. 147

Kessler, Kilian 134, 139/1–2, 141

Kokkinis, D., Schaffner, M., Studart, A.R., "Multimaterial magnetically assisted 3D printing of composite materials", Nat Commun, 6 8643 (2015). (Reprinted with permission) 166/1–3

Libanori, R., Erb, R.M., Studart, A.R., "Mechanics of platelet-reinforced composites assembled using mechanical and magnetic stimuli", ACS Appl. Mater. Interfaces, 5 (21) 10794-10805 (2013). Copyright 2013 American Chemical Society. (Reprinted with permission) 158

Living Architecture (LIAR) (illustrated by Sophie Nash) 169

Minke, Gernot (2012). Building with Bamboo, Design and Technology of a Sustainable Architecture 55

Moravánszky, Ákos 29/1

MOSO International BV 87/1–4, 89

MycoWorks Inc. 140

Natalio, Filipe (illustrated by Sophie Nash) 149; 150/2–3, 151/1–3; (illustrated by Luis Favas) 153; 154

National Geographic Creative/Alamy Stock Photo 80

Novarc Images/Alamy Stock Foto 168

Olsson, Johan (copyright Innventia) 96, 97, 124, 126/1–2, 127, 128, 129/1–2, 130, 131/1–2

Professorship Dirk E. Hebel (Carlina Teteris) Cover, 17/2, 46, 47, 48, 56/1, 56/2, 57, 58, 61/2, 62/1–2, 62/1–6, 64/1–3, 65/1–2, 66/1–2, 67/1–3, 69/1–2, 108

Professorship Dirk E. Hebel (Felix Heisel) 11/1; based on information by the Food and Agriculture Organization of the United Nations 33, 83; 34/1; based on information provided by Gandrea and Delboy 41/1; based on information provided by National Geographic 49; based on information provided by Oscar Hidalgo-Lopez (2003). Bamboo – The Gift of the Gods 60; 73/1–2; 162/1

Professorship Dirk E. Hebel (Jon Etter) 10/2

Professorship Dirk E. Hebel (Marta H. Wisniewska) 100, 104/1–2, 105, 145/2, 146/1–2, 148

Professorship Dirk E. Hebel (Sophie Nash) 11/2, 14; based on information provided by Oscar Hidalgo-Lopez: Bamboo – The Gift of the Gods 50, 51; based on information provided by Khosrow Ghavami 59/1; based on information provided by Felix Heisel and Felix Böck 76; based on information provided by the Pablo van der Lugt 94; 101, 109; based on information provided by Leon van Paassen 113; 117; based on information by ITKE University of Stuttgart 119/1; 125, 135; based on information provided by Henk M. Jonkers 143; based on information provided by Davide Carnelli 161/1, 162/2, 163/2–3, 164, 165

Professorship Dirk E. Hebel (Wojciech Zawarski) 19, 145/3

Ruvu Plantation (Tanzania Forest Service) and International Network of Bamboo and Rattan INBAR 82

Seitaridis, Elena 6–7, 23

shutterstock/Shebeko Cover

Siemens AG 68

Smithsonian Institution, National Museum of American History, Armed Forces History (Neg. No. 62468) 10/1

Stefano Paterna/Alamy Stock Foto 45/2

Stock.com/stockstudioX 53

Studart, André R. adapted from Ashby, M.F., Materials Selection in Mechanical Design Butterworth-Heinemann: Oxford, UK, 2005 163/1

Thijssen, Arjan/Materials and Environment, Delft University of Technology 144, 145/1

Van den Hoek, Sophia/DUS Architects 25

Van der Lugt, Pablo 86; based on information provided by the Food and Agriculture Organization FAO of the United Nations 90/1; (illustrated by Felix Heisel) 91; 90/2, 93/1–2

Van Paassen, Leon 110, 111, 112/1–2

Van Paassen, Leon/Visser & Smit Hanab 114/13

Völcker, Eckhard (www.voelcker.com) 61/1

weltspiegel 29/2

Wenig, Charlett 152

Yiping, Lou/International Network for Bamboo and Rattan INBAR 92

ZRS Architekten Ingenieure 41/2, 42, 43, 44/1–2

Index of Persons, Firms, and Institutions

Index of Projects, Buildings, and Places